INERT GAS SYSTEMS

1990 Edition

London, 1990

First published in 1982
by the INTERNATIONAL MARITIME ORGANIZATION
4 Albert Embankment, London SE1 7SR

Second edition 1983
Third edition 1990

Printed in the United Kingdom by CPI Books Limited, Reading RG1 8EX

ISBN 978-92-801-1262-7

IMO PUBLICATION
Sales number: I860E

C38580

FOREWORD

1 This publication contains the text of guidelines for inert gas systems and relevant IMO documents on inert gas systems and supersedes the publication 860 83.15.E. Provisions of the SOLAS Convention covering application and technical requirements for inert gas systems, together with recent developments on regulations for inert gas systems on chemical tankers are included with a view to setting out the framework as well as details of international requirements for inert gas systems.

Part I: Guidelines for Inert Gas Systems

2 The International Conference on Tanker Safety and Pollution Prevention, 1978, with resolution 5, requested the Organization to re-examine the requirements relating to inert gas systems in regulation II-2/62 of the 1974 SOLAS Convention, and to develop guidelines to supplement the requirements of that regulation.

3 The Maritime Safety Committee at its forty-second session approved the Guidelines for Inert Gas Systems (MSC/Circ.282), and at its forty-eighth session adopted amendments to sections 8 and 12 of the Guidelines (MSC/Circ.353). Furthermore, the Maritime Safety Committee at its fiftieth session adopted revised section 5.5 – cargo discharge – of the revised Guidelines (MSC/Circ.387). This part provides the text of the Guidelines for Inert Gas Systems which incorporates all these amendments.

Part II: Provisions of application

4 Provisions of application for inert gas systems in the 1974 SOLAS Convention were modified by the 1978 SOLAS Protocol and amended by the 1981 and 1983 amendments. This part contains all the relevant provisions of application as well as clarification of application of inert gas system requirements developed by the Maritime Safety Committee at its fifty-fifth session (MSC/Circ.485).

Part III: Provisions of technical requirements

5 Technical requirements for inert gas systems in the 1974 SOLAS Convention were amended substantially by the 1981 amendments and to a lesser extent by the 1983 amendments. These technical requirements which are to be applied under the provisions of the SOLAS Convention are contained in this part.

Part IV: Regulations for inert gas systems on chemical tankers

6 Regulation II-2/60 of the 1978 SOLAS Protocol requires new and existing chemical tankers of a certain size, when carrying petroleum products, to be fitted with a fixed inert gas system at specific dates. The Assembly at its twelfth session, recognizing the unique features of chemical tankers, adopted the Interim regulation

for inert gas systems on chemical tankers carrying petroleum products (resolution A.473(XII)), and urged Governments concerned, under the provisions of regulation I/5 of the 1974 SOLAS Convention, to apply to chemical tankers as appropriate the Interim regulation as equivalent to regulation II-2/62 of the 1974 SOLAS Convention.

7 The Assembly at its fourteenth session, recognizing that the extension of the regulation in resolution A.473(XII) to cover the carriage of petroleum and other liquid products would be desirable, adopted resolution A.567(14), the regulation for inert gas systems on chemical tankers, which supersedes resolution A.473(XII).

8 The Assembly at its fourteenth session also adopted draft amendments to regulation II-2/55.5 of the International Convention for the Safety of Life at Sea, 1974, as amended (resolution A.566(14)). These amendments were subsequently adopted by the Maritime Safety Committee at its fifty-seventh session and included in the 1989 set of amendments to the SOLAS Convention expected to enter into force on 1 February 1992 and the resolution is also contained in this part.

Part V: Application of requirements for inert gas systems for oil tankers by port authorities and terminal operators (MSC/Circ.329)

9 At the forty-sixth session of the Maritime Safety Committee, concern was expressed on the stringent oxygen levels insisted on by some terminal operators and port authorities for inerted cargo tanks of oil tankers, and their reluctance to allow the opening of inerted tanks for dipping, measuring and sampling. The Committee, in MSC/Circ.329, noting the concern, urged Governments to encourage port authorities and terminal operators to comply with international requirements.

CONTENTS

PART I GUIDELINES FOR INERT GAS SYSTEMS

(adopted by the Maritime Safety Committee at its forty-second session and amended at its forty-eighth and fiftieth sessions)

CONTENTS

1 INTRODUCTION

1.1 Purpose

The International Conference on Tanker Safety and Pollution Prevention held in February 1978 passed resolution 5 recommending that the International Maritime Organization develop Guidelines to supplement the requirements of amended regulation 62 of chapter II-2 of the 1974 SOLAS Convention* by taking into account the arduous operating conditions of inert gas systems and the need to maintain them to a satisfactory standard. In addition regulation 62.1 requires that an inert gas system shall be designed, constructed and tested to the satisfaction of the Administration. These Guidelines have accordingly been developed to supplement and complement the Convention requirements for inert gas systems. They are offered to Administrations to assist them in determining appropriate design and constructional parameters and in formulating suitable operational procedures when inert gas systems are installed in ships flying the flag of their State.

1.2 Application

1.2.1 The status of these Guidelines is advisory. They are intended to cover the design and operation of:

.1 inert gas systems that are required on new tankers by regulation 60 of chapter II-2 of the 1978 SOLAS Protocol and in accordance with regulation 62;

.2 inert gas systems that are required on existing tankers by regulation 60 of chapter II-2 of the 1978 SOLAS Protocol and in accordance with regulation 62.20;

.3 inert gas systems which are fitted but not required to comply with the requirements of regulation 60 of chapter II-2 of the 1978 SOLAS Protocol.

1.2.2 However, for existing inert gas systems the Guidelines are directed primarily at operational procedures and are not intended to be interpreted as requiring modifications to existing equipment other than those which are required on ships to which regulation 62.20 applies.

1.2.3 The content of these Guidelines is based on current general practice used in the design and operation of inert gas systems using flue gas from the uptake from the ship's main or auxiliary boilers, and installed on crude oil tankers and combination carriers. The Guidelines do not exclude other sources of inert gas, such as systems incorporating independent inert gas generators, other designs, materials or operational procedures. All such divergences should be carefully assessed to ensure that they achieve the objectives of these Guidelines.

---

* Any reference to regulation 62 in these Guidelines means the new text of regulation 62 of chapter II-2 of the 1983 SOLAS amendments, as adopted by the Maritime Safety Committee at its forty-eighth session in June 1983.

1.3 Definitions

1.3.1 *Inert gas* means a gas or a mixture of gases, such as flue gas, containing insufficient oxygen to support the combustion of hydrocarbons.

1.3.2 *Inert condition* means a condition in which the oxygen content throughout the atmosphere of a tank has been reduced to 8% or less by volume by addition of inert gas.

1.3.3 *Inert gas plant* means all equipment specially fitted to supply, cool, clean, pressurize, monitor and control delivery of inert gas to cargo tank systems.

1.3.4 *Inert gas distribution system* means all piping, valves, and associated fittings to distribute inert gas from the inert gas plant to cargo tanks, to vent gases to atmosphere and to protect tanks against excessive pressure or vacuum.

1.3.5 *Inert gas system* means an inert gas plant and inert gas distribution system together with means for preventing backflow of cargo gases to the machinery spaces, fixed and portable measuring instruments and control devices.

1.3.6 *Inerting* means the introduction of inert gas into a tank with the object of attaining the inert condition defined in 1.3.2.

1.3.7 *Gas-freeing* means the introduction of fresh air into a tank with the object of removing toxic, flammable and inert gases and increasing the oxygen content to 21% by volume.

1.3.8 *Purging* means the introduction of inert gas into a tank already in the inert condition with the object of:

.1 further reducing the existing oxygen content; and/or

.2 reducing the existing hydrocarbon gas content to a level below which combustion cannot be supported if air is subsequently introduced into the tank.

1.3.9 *Topping up* means the introduction of inert gas into a tank which is already in the inert condition with the object of raising the tank pressure to prevent any ingress of air.

2 PRINCIPLES

2.1 General

With an inert gas system the protection against a tank explosion is achieved by introducing inert gas into the tank to keep the oxygen content low and reduce to safe proportions the hydrocarbon gas concentration of the tank atmosphere.

2.2 Flammable limits

2.2.1 A mixture of hydrocarbon gas and air cannot ignite unless its composition lies within a range of gas in air concentrations known as the "flammable range". The lower limit of this range, known as the "lower flammable limit" is any hydrocarbon concentration below which there is insufficient hydrocarbon gas to support combustion. The upper limit of the range, known as the "upper flammable limit" is any hydrocarbon concentration above which there is insufficient air to support combustion.

2.2.2 The flammable limits vary somewhat for different pure hydrocarbon gases and for the gas mixtures derived from different petroleum liquids. In practice, however, the lower and upper flammable limits of oil cargoes carried in tankers can be taken, for general purposes, to be 1% and 10% hydrocarbon by volume, respectively.

2.3 Effect of inert gas on flammability

2.3.1 When an inert gas is added to a hydrocarbon gas/air mixture the result is to increase the lower flammable limit concentration and to decrease the upper flammable limit concentration. These effects are illustrated in figure 1, which should be regarded only as a guide to the principles involved.

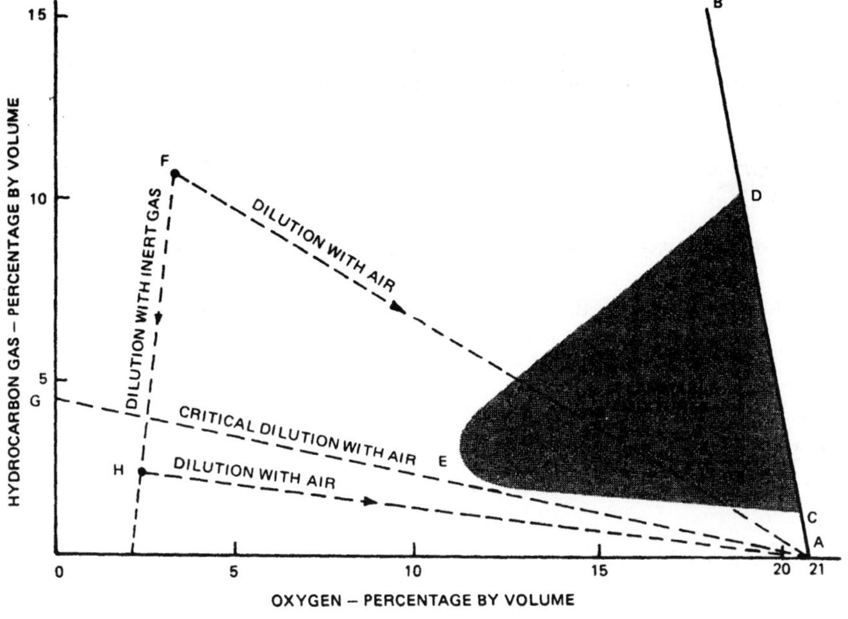

Figure 1

2.3.2 Any point on the diagram represents a hydrocarbon gas/air/inert gas mixture, specified in terms of its hydrocarbon and oxygen content. Hydrocarbon/air mixtures without inert gas lie on the line AB, the slope of which shows the reduction in oxygen content as the hydrocarbon content increases. Points to the left of AB represent mixtures with their oxygen content further reduced by the addition of inert gas. It is evident from figure 1 that as inert gas is added to hydrocarbon/air mixtures the flammable range progressively decreases until the oxygen content reaches a level generally taken to be about 11% by volume, at which no mixture can burn. The figure of 8% by volume specified in these Guidelines for a safely inerted gas mixture allows some margin beyond this value.

2.3.3 The lower and upper flammability limit mixtures for hydrocarbon gas in air are represented by the points C and D. As the inert gas content increases, the flammable limit mixtures change. This is indicated by the lines CE and DE, which finally converge at the point E. Only those mixtures represented by points in the shaded area within the loop CED are capable of burning. Changes of composition, due to the addition of either air or inert gas, are represented by movements along straight lines. These lines are directed either towards the point A (pure air), or towards a point on the oxygen content axis corresponding to the composition of the added inert gas. Such lines are shown for the gas mixture represented by the point F.

2.3.4 When an inert mixture, such as that represented by the point F, is diluted by air its composition moves along the line FA and therefore enters the shaded area of flammable mixtures. This means that all inert mixtures in the region above the line GA (critical dilution line) pass through a flammable condition as they are mixed with air (for example during a gas-freeing operation). Those below the line GA, such as that represented by point H, do not become flammable on dilution. It will be noted that it is possible to move from a mixture, such as that represented by F, to one such as that represented by H, by dilution with additional inert gas, i.e. purging.

2.4 Sources

Possible sources of inert gas on tankers including combination carriers are:

.1 the uptake from the ship's main or auxiliary boilers;

.2 an independent inert gas generator; or

.3 a gas turbine plant when equipped with an afterburner.

2.5 Quality

Good combustion control in the ship's boilers is necessary to achieve an oxygen content of 5% by volume. In order to obtain this quality, it may be necessary to use automatic combustion control.

2.6 Methods of gas replacement

2.6.1 There are three operations which involve replacement of gas in cargo tanks, namely:

.1 inerting;

.2 purging;

.3 gas-freeing.

2.6.2 In each of these replacement operations, one of two processes can predominate:

.1 dilution, which is a mixing process (see 2.6.3);

.2 displacement, which is a layering process (see 2.6.4).

These two processes have a marked effect on the method of monitoring the tank atmosphere and the interpretation of the results. Figures 3 and 5 show that an understanding of the nature of the gas replacement process actually taking place within the tank is necessary for the correct interpretation of the reading shown on the appropriate gas sampling instrument.

2.6.3 The dilution theory assumes that the incoming gas mixes with the original gases to form a homogeneous mixture throughout the tank. The result is that the concentration of the original gas decreases exponentially. In practice the actual rate of gas replacement depends upon the volume flow of the incoming gas, its entry velocity, and the dimensions of the tank. For complete gas replacement it is important that the entry velocity of the incoming gas is high enough for the jet to reach the bottom of the tank. It is therefore important to confirm the ability of every installation using this principle to achieve the required degree of gas replacement throughout the tank.

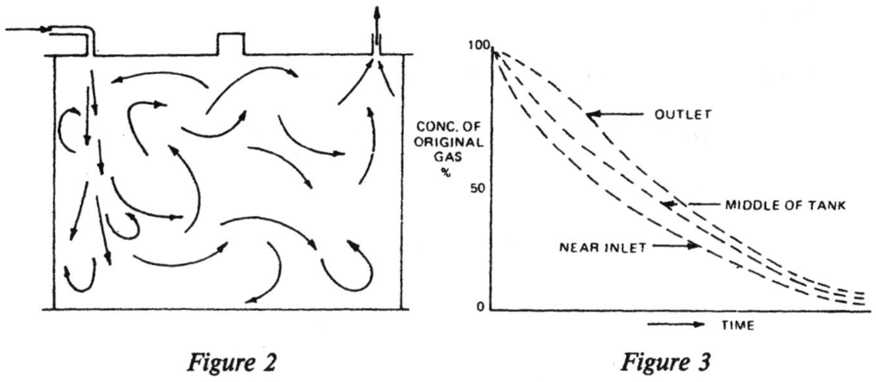

Figure 2 *Figure 3*

Figure 2 shows an inlet and outlet configuration for the dilution process and illustrates the turbulent nature of the gas flow within the tank.

Figure 3 shows typical curves of gas concentration against time for three different sampling positions.

2.6.4 Ideal replacement requires a stable horizontal interface between the lighter gas entering at the top of the tank and the heavier gas being displaced from the bottom of the tank through some suitable piping arrangement. This method requires a relatively low entry velocity of gas and in practice more than one volume change is necessary. It is therefore important to confirm the ability of every installation using this principle to achieve the required degree of gas replacement throughout the tank.

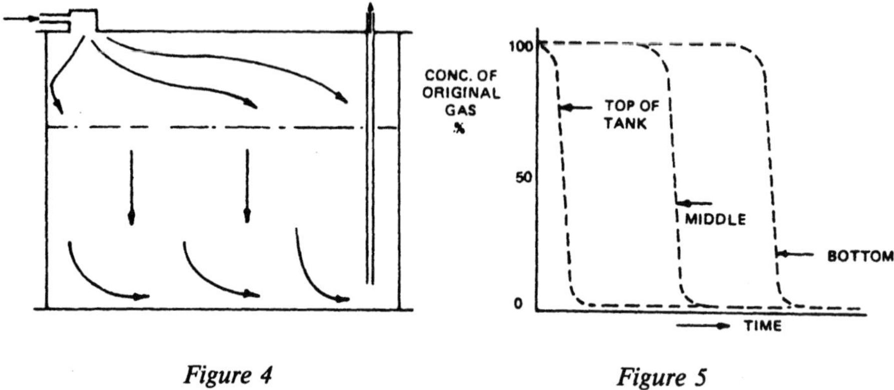

Figure 4 *Figure 5*

Figure 4 shows an inlet and outlet configuration for the displacement process, and indicates the interface between the incoming and outgoing gases.

Figure 5 shows typical curves of gas concentration against time for three different sampling levels.

2.7 General policy of cargo tank atmosphere control

2.7.1 Tankers fitted with an inert gas system should have their cargo tanks kept in a nonflammable condition at all times (see figure 1). It follows that:

.1 tanks should be kept in the inert condition whenever they contain cargo residues or ballast. The oxygen content should be kept at 8% or less by volume with a positive gas pressure in all the cargo tanks;

.2 the atmosphere within the tank should make the transition from the inert condition to the gas-free condition without passing through the flammable condition. In practice this means that before any tank is gas-freed, it would be purged with inert gas until the hydrocarbon content of the tank atmosphere is below the critical dilution line (see figure 1);

.3 when a ship is in a gas-free condition before arrival at a loading port, tanks should be inerted prior to loading.

2.7.2 In order to maintain cargo tanks in a nonflammable condition the inert gas plant will be required to:

.1 inert empty cargo tanks (see 5.1);

.2 be operated during cargo discharge, deballasting and necessary in-tank operations (see 5.2, 5.5, 5.6, 5.8 and 5.9);

.3 purge tanks prior to gas-freeing (see 5.10);

.4 top up pressure in the cargo tanks when necessary, during other stages of the voyage (see 5.4 and 5.8).

3 FUNCTION AND DESIGN CONSIDERATIONS

This section addresses itself to inert flue gas systems. The design of systems other than inert flue gas systems should take into account, whenever applicable, the general principles outlined in this section.

3.1 Description of an inert flue gas system

3.1.1 A typical arrangement for an inert flue gas system is shown in figure 6. It consists of flue gas isolating valves located at the boiler uptake points through which pass hot, dirty gases to the scrubber and demister. Here the gas is cooled and cleaned before being piped to blowers which deliver the gas through the deck water seal, the nonreturn valve, and the deck isolating valve to the cargo tanks. A gas pressure regulating valve is fitted downstream of the blowers to regulate the flow of gases to the cargo tank. A liquid-filled pressure/vacuum breaker is fitted to prevent excessive pressure or vacuum from causing structural damage to cargo tanks. A vent is fitted between the deck isolating/nonreturn valve and the gas pressure regulating valve to vent any leakage when the plant is shut down.

3.1.2 For delivering inert gas to the cargo tanks during cargo discharge, deballasting, tank cleaning and for topping up the pressure of gas in the tank during other phases of the voyage, an inert gas deck main runs forward from the deck isolating valve for the length of the cargo deck. From this inert gas main, inert gas branch lines lead to the top of each cargo tank.

3.2 Function of inert gas scrubber

3.2.1 The purpose of the scrubber is to cool the flue gas and remove most of the sulphur dioxide and particulate soot. All three actions are achieved by direct contact between the flue gas and large quantities of seawater.

3.2.2 Before entering the bottom of the scrubbing tower the gas is cooled by being passed either through a water spray, or bubbled through a water seal. Such a seal may also serve as the additional safety device to prevent any leakage of gas from the boiler uptake when the scrubber is opened up for inspection or maintenance.

3.2.3 In the scrubbing tower itself the gas moves upwards through downward flowing water. For maximum contact between gas and water, several layers made up of one or more of the following arrangements may be fitted:

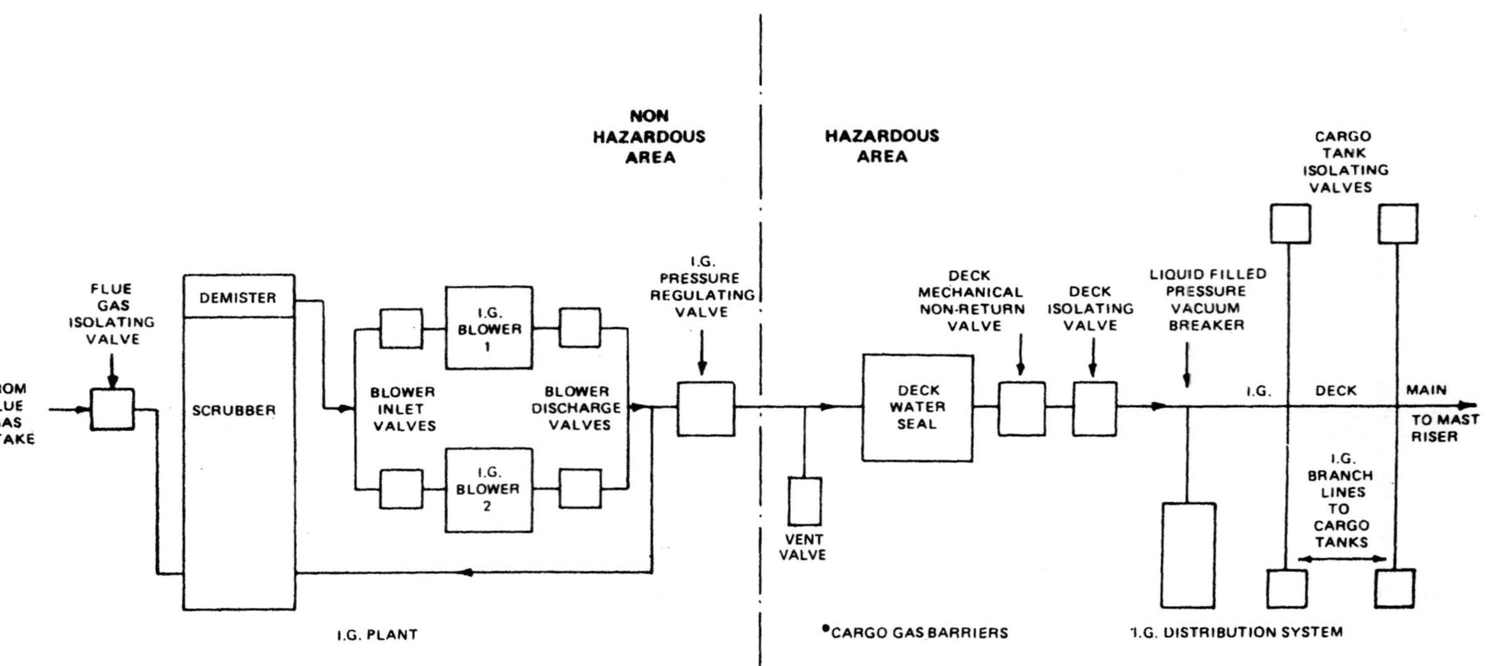

Figure 6 A typical arrangement for an inert gas system

.1 spray nozzles;

.2 trays of "packed" stones or plastic chippings;

.3 perforated "impingement" plates;

.4 venturi nozzles and slots.

3.2.4 At the top of or downstream of the scrubbing tower, water droplets are removed by one or more demisters which may be polypropylene mattresses or cyclone dryers.

Designs of individual manufacturers vary considerably.

3.3 Design considerations for inert gas scrubber

3.3.1 The scrubber should be of a design related to the type of tanker, cargoes and combustion control equipment of the inert gas supply source and be capable of dealing with the quantity of inert gas required by regulation 62 at the designed pressure differential of the system.

3.3.2 The performance of the scrubber at full gas flow should be such as to remove at least 90% of sulphur dioxide and to remove solids effectively. In product carriers more stringent requirements may be needed for product quality.

3.3.3 The internal parts of the scrubber should be constructed in corrosion-resistant materials in respect of the corrosive effect of the gas. Alternatively, the internal parts may be lined with rubber, glass fibre epoxy resin or other equivalent material, in which case the flue gases may require to be cooled before they are introduced into the lined sections of the scrubber.

3.3.4 Adequate openings and sight glasses should be provided in the shell for inspection, cleaning and observational purposes. The sight glasses should be reinforced to withstand impact and be of a heat resisting type. This may be achieved by the use of double glazing.

3.3.5 The design should be such that under normal conditions of trim and list the scrubber efficiency will not fall by more than 3%, nor will the temperature rise at the gas outlet exceed the designed gas outlet temperature by more than 3°C.

3.3.6 The location of the scrubber above the load waterline should be such that the drainage of the effluent is not impaired when the ship is in the fully loaded condition.

3.4 Function of inert gas blowers

3.4.1 Blowers are used to deliver the scrubbed flue gas to the cargo tanks. Regulation 62.3.1 requires that at least two blowers shall be provided which together shall be capable of delivering inert gas to the cargo tanks at a rate of at least 125% of the maximum rate of discharge capacity of the ship expressed as a volume.

3.4.2 In practice, installations vary from those which have one large blower and one small blower, whose combined total capacity complies with regulation 62, to those in which each blower can meet this requirement. The advantage claimed for the former is that it is convenient to use a small capacity blower when topping up the gas pressure in the cargo tanks at sea; the advantage claimed for the latter is that if either blower is defective the other one is capable of maintaining a positive gas pressure in the cargo tanks without extending the duration of the cargo discharge.

3.5 Design considerations for inert gas blowers

3.5.1 The blower casing should be constructed in corrosion-resistant material or alternatively of mild steel but then its internal surfaces should be stove-coated, or lined with rubber or glass fibre epoxy resin or other equivalent material to protect it from the corrosive effect of the gas.

3.5.2 The impellers should be manufactured in a corrosion-resistant material. Aluminium bronze impellers should be stress-relieved after welding. All impellers should be tested by overspeeding to 20% above the design running speed of the electric motor or 10% above the speed at which the overspeed trip of the turbine would operate, whichever is applicable.

3.5.3 Substantial drains, fitted with adequate water seals, should be provided in the casing to prevent damage by an accumulation of water. The drains should be in accordance with the provisions of 3.15.4.

3.5.4 Means should be provided such as fresh water washing to remove the build-up of deposits which would cause vibration during blower operation.

3.5.5 The casing should be adequately ribbed to prevent panting and should be so designed and arranged as to facilitate the removal of the rotor without disturbing major parts of the inlet and outlet gas connections.

3.5.6 Sufficient openings in the casing should be provided to permit inspection.

3.5.7 Where separate shafts are provided for the prime mover and the blower, a flexible coupling between these shafts should be provided.

3.5.8 When roller or ball bearings are used, due regard should be paid to the problem of brinelling and the method of lubrication. The type of lubrication chosen, i.e. oil or grease, should have regard to the diameter and rotational speed of the shaft. If sleeve bearings are fitted then resilient mountings are not recommended.

3.5.9 The blower pressure/volume characteristics should be matched to the maximum system requirements. The characteristics should be such that in the event of the discharge of any combination of cargo tanks at the discharge rate indicated in 3.4, a minimum pressure of 200 mm water gauge is maintained in any cargo tank after allowing for pressure losses due to:

.1 the scrubber tower and demister;

.2 the piping conveying the hot gas to the scrubbing tower;

.3 the distribution piping downstream of the scrubber;

.4 the deck water seal;

.5 the length and diameter of the inert gas distribution system.

3.5.10 When both blowers are not of equal capacity the pressure/volume characteristics and inlet and outlet piping should be so matched that if both blowers can be run in parallel, they are able to develop their designed outputs. The arrangements should be such as to prevent the blower on load from motoring the blower that is stopped or has tripped out.

3.5.11 If the prime mover is an electric motor then it should be of sufficient power to ensure that it will not be overloaded under all possible operating conditions of the blower. The overload power requirement should be based on the blower inlet conditions of −5°C at −400 mm water gauge and outlet conditions of 0°C and atmospheric pressure. Arrangements should be provided, if necessary, to maintain the windings in a dry condition during the inoperative period.

3.6 Function of nonreturn devices

3.6.1 The deck water seal and mechanical nonreturn valve together form the means of automatically preventing the backflow of cargo gases from the cargo tanks to the machinery space or other safe area in which the inert gas plant is located.

3.6.2 *Deck water seal (see regulation 62.10)*

This is the principal barrier. A water seal is fitted which permits inert gas to be delivered to the deck main but prevents any backflow of cargo gas even when the inert gas plant is shut down. It is vital that a supply of water is maintained to the seal at all times, particularly when the inert gas plant is shut down. In addition, drains should be led directly overboard and should not pass through the machinery spaces. There are different designs but one of three principal types may be adopted.

.1 *Wet type*

This is the simplest type of water seal. When the inert gas plant is operating, the gas bubbles through the water from the submerged inert gas inlet pipe, but if the tank pressure exceeds the pressure in the inert gas inlet line the water is pressed up into this inlet pipe and thus prevents backflow. The drawback of this type of water seal is that water droplets may be carried over with the inert gas which, although it does not impair the quality of the inert gas, could increase corrosion. A demister should, therefore, be fitted in the gas outlet from the water seal to reduce any carry-over. Figure 7 shows an example of this type.

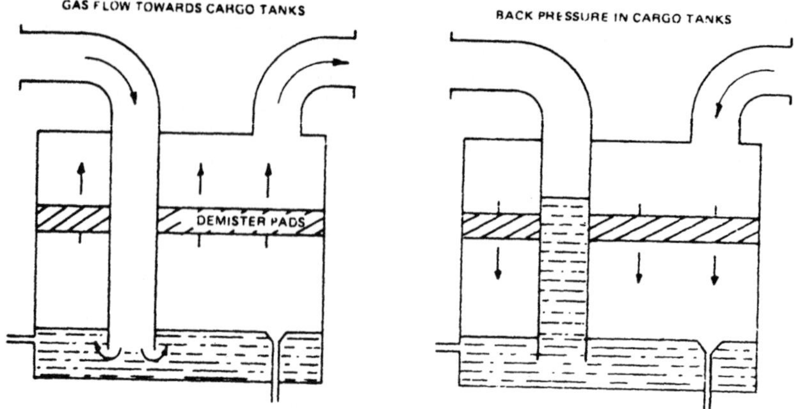

Figure 7 Deck water seal - wet type.

.2 *Semi-dry type*

Instead of bubbling through the water trap the inert gas flow draws the sealing water into a separate holding chamber by venturi action thus avoiding or at least reducing the amount of water droplets being carried over. Otherwise it is functionally the same as wet type. Figure 8 shows an example of this type.

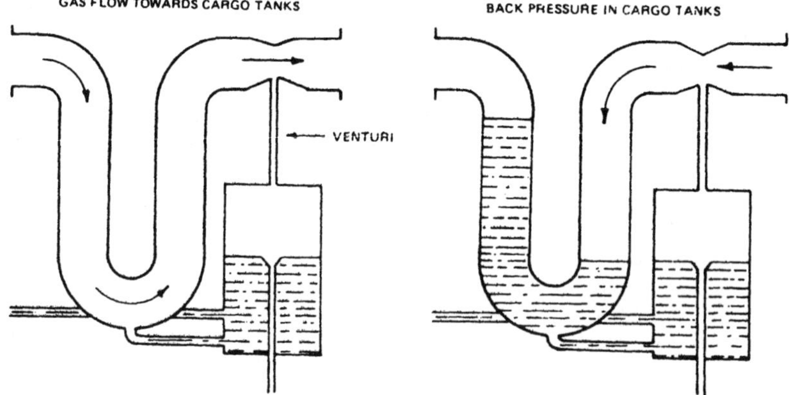

Figure 8 Deck water seal - semi-dry type.

.3 *Dry type*

In this type the water is drained when the inert gas plant is in operation (gas flowing to the tanks) and filled with water when the inert gas plant is either shut down or the tank pressure exceeds the inert gas blower discharge pressure. Filling and drainage are performed by automatically operated valves controlled by the levels in the water seal and drop tanks and by the operating state of the blowers. The advantage of this type is that water carry-over is prevented. The drawback could be the risk of failure of the automatically controlled valves which may render the water seal ineffective. Figure 9 shows an example of this type.

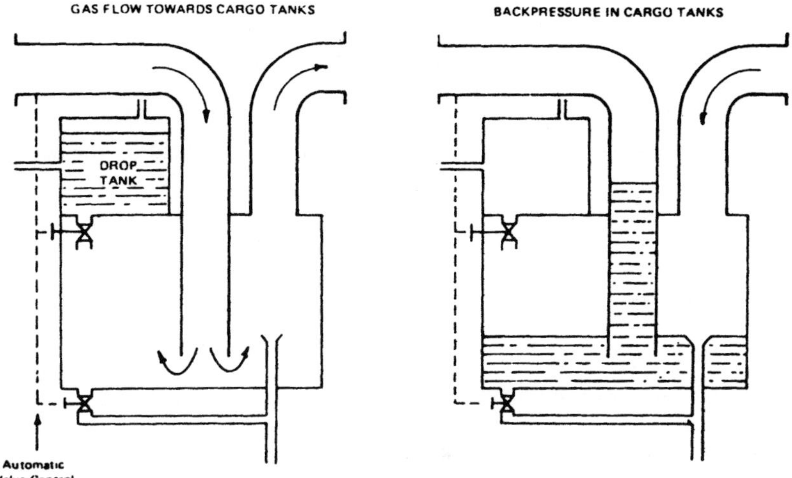

Figure 9 Deck water seal – dry type.

3.6.3 *Deck mechanical nonreturn valve and deck isolating valve*

As a further precaution to avoid any backflow of gas from the cargo tanks, and to prevent any backflow of liquid which may enter the inert gas main if the cargo tanks are overfilled, regulation 62.10.8 requires a mechanical nonreturn valve, or equivalent, which should be fitted forward of the deck water seal and should operate automatically at all times.

This valve should be provided with a positive means of closure or, alternatively, a separate deck isolating valve fitted forward of the nonreturn valve, so that the inert gas deck main may be isolated from the nonreturn devices. The separate isolating valve has the advantage of enabling maintenance work to be carried out on the nonreturn valve.

3.6.4 *Inert gas valve (see regulation 62.10.9)*

This valve should be opened when the inert gas plant is shut down to prevent leakage past the nonreturn devices from building up any pressure in the inert gas line between the gas pressure regulating valve and these nonreturn devices.

3.7 Design considerations for nonreturn devices

3.7.1 The material used in the construction of the nonreturn devices should be resistant to fire and to the corrosive attack from acids formed by the gas. Alternatively low carbon steel protected by a rubber lining or coated with glass fibre epoxy resin or equivalent material may be used. Particular attention should be paid to the gas inlet pipe to the water seal.

3.7.2 The deck water seal should present a resistance to backflow of not less than the pressure setting of the pressure/vacuum breaking device on the inert gas distribution system and should be so designed as to prevent the backflow of gases under any foreseeable operating conditions.

3.7.3 The water in the deck seal should be maintained by a regulating flow of clean water through the deck seal reservoir.

3.7.4 Sight glasses and inspection openings should be provided on the deck seal to permit satisfactory observation of the water level during its operation and to facilitate a thorough survey. The sight glasses should be reinforced to withstand impact.

3.7.5 Any drains from the nonreturn devices should incorporate a water seal in accordance with 3.15.4 and should comply generally with 3.16.

3.8 Inert gas distribution system

3.8.1 The inert gas distribution system, together with the cargo tank venting system, where applicable, has to provide:

.1 means of delivering inert gas to the cargo tanks during discharge, deballasting and tank cleaning operations, and for topping up the pressure of gas in the tank;

.2 means of venting tank gases to atmosphere during cargo loading and ballasting;

.3 additional inlet or outlet points for inerting, purging and gas-freeing;

.4 means of isolating individual tanks from the inert gas main for gas-freeing (see 3.12.4);

.5 means of protecting tanks from excessive pressure or vacuum.

3.8.2 A large variety of designs and operational procedures may be used to meet these interrelated requirements. In 3.9 are considered some of the major design options and their more important operational consequences; further advice on operational precautions is given in section 5.

3.9 Design considerations for valves and pipework in inert gas systems

3.9.1 The flue gas uptake point should be so selected that the gas is not too hot for the scrubber, nor causes hard deposits on the flue gas isolating valves. It should not be so close to the uptake outlet that air can be drawn into the system. When boilers are fitted with rotary air heaters, the offtake point should be before the air heater inlet.

3.9.2 The materials used for flue gas isolating valves should take into account the temperature of gas at the offtake. Cast iron is acceptable for temperatures below 220°C. Valves exposed to a temperature exceeding 220°C should be made from a material not only compatible with the temperature but also resistant to the corrosive effect of stagnant flue gases.

3.9.3 Flue gas isolating valves should be provided with facilities to keep the seatings clear of soot unless the valve is designed to close with a seat cleaning action. Flue gas isolating valves should also be provided with air sealing arrangements.

3.9.4 If expansion bellows are considered necessary they should have a smooth internal sleeve and preferably be mounted so that the gas flow through them is vertical. They should be constructed of material resistant to stagnant damp acidic soot.

3.9.5 The pipework between the flue gas isolating valve and the scrubber should be made from heavy gauge steel resistant to corrosion and so arranged as to prevent the accumulation of damp acidic soot by the avoidance of unnecessary bends and branches.

3.9.6 The inlet piping to the scrubber should be so arranged as to permit positive isolation from the flue gases prior to gas-freeing the scrubber for entry for maintenance purposes. This may be accomplished by the removal of a suitable length of pipe section and blanking, by spectacle flanges or by a water seal which would prevent any leakage of gas from the boiler when the flue gas isolating valve is shut and the scrubber is opened up for inspection and maintenance. In the event that the drainage of the water seal is itself required for inspection purposes, then isolation should be achieved either by removal of the suitable lengths of pipe sections and blanking, or by the use of spectacle flanges.

3.9.7 The gas outlet piping from the scrubber to the blowers and recirculating lines should be made from steel suitably coated internally.

3.9.8 Suitable isolating arrangements should be incorporated in the inlet and outlet of each blower to permit safe overhaul and maintenance of a blower while permitting the use of the inert gas system using the other blower.

3.9.9 The regulating valve required by regulation 62.9.1 should be provided with means to indicate whether the valve is open or shut. Where the valve is used to regulate the flow of inert gas it should be controlled by the inert gas pressure sensed between the deck isolating valve and the cargo tanks.

3.9.10 Deck lines should be of steel and be so arranged as to be self draining and should be firmly attached to the ship's structure with suitable arrangements to take into account movement due to heavy weather, thermal expansion and flexing of the ship.

3.9.11 The diameter of the inert gas main, valves and branch pipes should take account of the system requirements detailed in 3.5.9. To avoid excessive pressure drop, the inert gas velocity should not exceed 40 m/s in any section of the distribution system when the inert gas system is operating at its maximum capacity. If the inert gas main is used for venting during loading, other factors may need to be taken into consideration as developed in regulation 58 of chapter II-2 of the 1974 SOLAS Convention* for cargo tank venting systems.

---

* Regulation 59 of chapter II-2 of the 1983 SOLAS amendments.

3.9.12 All pressure and vacuum relief openings should be fitted with flame screens with easy access for cleaning and renewal. The flame screens should be at the inlets and outlets of any relief device and be of robust construction sufficient to withstand the pressure of gas generated at maximum loading and during ballasting operations while presenting minimum resistance.

3.10 Gas pressure regulating valves and recirculating arrangements

3.10.1 Pressure control arrangements should be fitted to fulfil two functions:

.1 to prevent automatically any backflow of gas in the event of either a failure of the inert gas blower, scrubber pump, etc., or when the inert gas plant is operating correctly but the deck water seal and mechanical nonreturn valve have failed and the pressure of gas in the tank exceeds the blower discharge pressure, e.g. during simultaneous stripping and ballasting operations;

.2 to regulate the flow of inert gas to the inert gas deck main.

3.10.2 A typical arrangement by which 3.10.1.2 can be achieved is as follows:

Systems with automatic pressure control and a gas recirculating line. These installations permit control of inert gas pressure in the deck main without having to adjust the inert gas blower speed. Gas not required in the cargo tanks is recirculated to the scrubber or vented to atmosphere. Gas pressure regulating valves are fitted in both the main and recirculating lines; one is controlled by a gas pressure transmitter and regulator, while the other may be controlled either in a similar manner or by a weight-operated valve. The pressure transmitter is sited downstream of the deck isolating valve; this enables a positive pressure to be maintained in the cargo tanks during discharge. However, it does not necessarily ensure that the scrubber is not overloaded during inerting and purging operations.

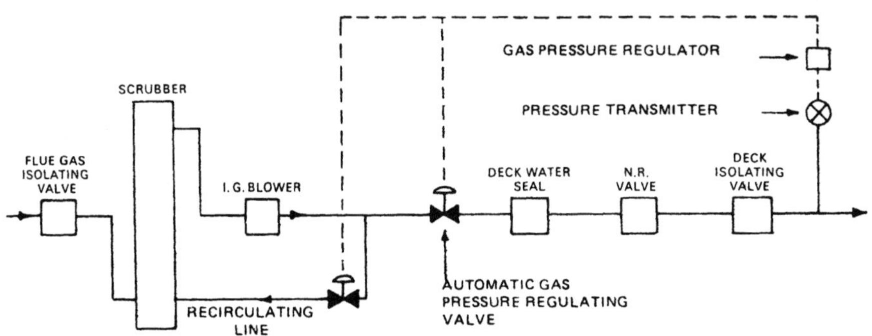

Figure 10 Typical automatic pressure control system.

Alternative methods which comply with regulation 62.9.1 may be considered.

3.11 **Arrangements for inerting, purging and gas-freeing** (regulation 62.13)

3.11.1 The principles of dilution and displacement have already been described in 2.6.3 and 2.6.4. Their application to specific installations depends on a variety of factors, including:

.1 the results of laboratory tests;

.2 whether or not purging of hydrocarbon gas is required in every tank on every voyage; and

.3 the method of venting cargo tank vapours.

3.11.2 Several arrangements are possible. One feature which should be common to all is that the inlet and outlet points should be so located that efficient gas replacement can take place throughout the tank.

There are three principal arrangements:

Arrangement	Inlet point	Outlet point	Principle
I	top	top	dilution
II	bottom	top	dilution
III	top	bottom	displacement or dilution

It will be noted that all three arrangements can be used for inerting, purging and gas-freeing, and that a particular ship design may incorporate more than one arrangement.

3.11.3 *Arrangement I*

Gases are both introduced and vented from the top of the tank. This is the simplest arrangement. Gas replacement is by the dilution method. The incoming gas should always enter the tank in such a way as to achieve maximum penetration and thorough mixing throughout the tank. Gases can be vented through a vent stack on each tank or through a common vent main. (See figure 11).

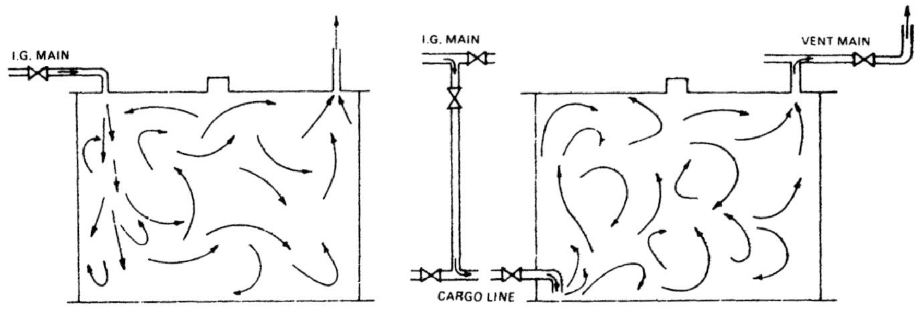

Figure 11 Dilution (I). *Figure 12* Dilution (II).

3.11.4 *Arrangement II*

Gas is introduced at the bottom of the tank and vented from the top. Gas replacement is by the dilution method. This arrangement introduces the gas through a connection between the inert gas deck main (just forward of the mechanical nonreturn valve) and the bottom cargo lines (see figure 12). A special fixed gas-freeing fan may also be fitted. Exhaust gas may be vented through individual vent stacks or, if valves are fitted to isolate each cargo tank from the inert gas main, through this main to the mast riser or high velocity vent.

3.11.5 *Arrangement III*

Gas is introduced at the top of the tank and discharged from the bottom. This arrangement permits the displacement method (see figure 13), although the dilution method may predominate if the density difference between the incoming and existing gases is small or the gas inlet velocity is high (see figure 14). The inert gas inlet point is often led horizontally into a tank hatch in order to minimize turbulence at the interface. The outlet point is often a specially fitted purge pipe extending from within 1 m of the bottom plating to 2 m above deck level (to minimize the amount of vapour at deck level.)

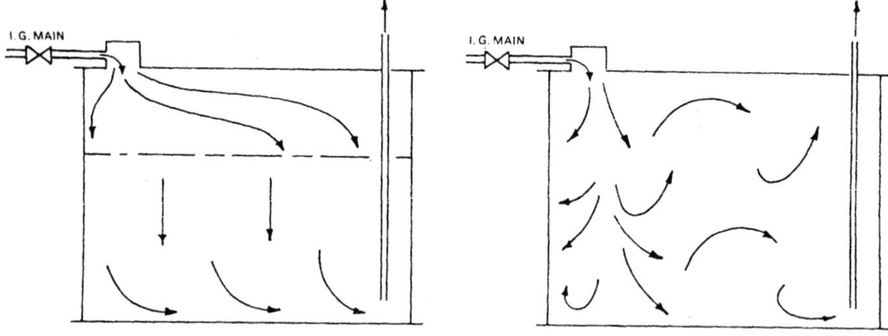

Figure 13 Displacement (III). *Figure 14* Dilution (III).

3.12 Isolation of cargo tanks from the inert gas deck main (regulation 62.11)

3.12.1 For gas-freeing and tank entry some valve or blanking arrangement is always fitted to isolate individual cargo tanks from the inert gas deck main.

3.12.2 The following factors should be considered in choosing a suitable arrangement:

.1 protection against gas leakage or incorrect operation during tank entry;

.2 ease and safety of use;

.3 facility to use the inert gas main for routine gas-freeing operations;

.4 facility to isolate tanks for short periods for the regulation of tank pressures and manual ullaging;

.5 protection against structural damage due to cargo pumping and ballasting operations when a cargo tank is inadvertently isolated from the inert gas main.

3.12.3 In no case should the arrangement prevent the proper venting of the tank.

3.12.4 Some examples of arrangements in use are shown in figure 15.

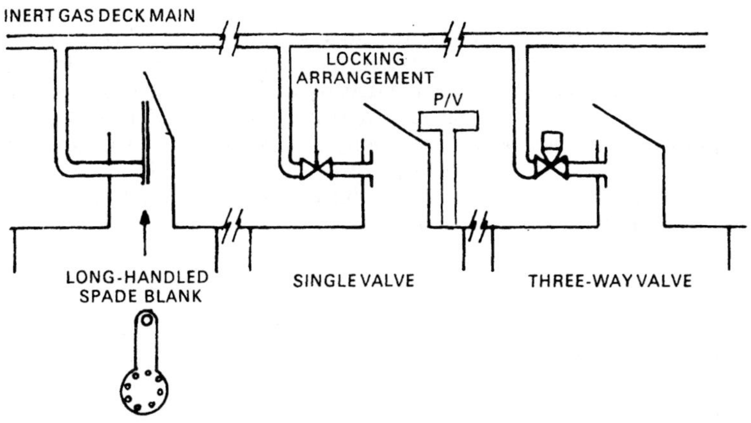

Figure 15 Examples of methods for isolating tanks from the inert gas main.

3.13 Liquid-filled pressure/vacuum breakers

3.13.1 One or more liquid-filled pressure/vacuum breakers should be fitted, unless pressure/vacuum valves having the capacity to prevent excessive pressure or vacuum in accordance with requirements of regulation 62.14 are fitted.

3.13.2 These devices require little maintenance, but will only operate at the required pressure if they are filled to the correct level with liquid of the correct density. Either a suitable oil or a freshwater/glycol mixture should be used to prevent freezing in cold weather. Evaporation, ingress of seawater, condensation and corrosion should be taken into consideration and adequately compensated for. In heavy weather, the pressure surge caused by the motion of liquid in the cargo tanks may cause the liquid of the pressure/vacuum breaker to be blown out (see figure 16).

3.13.3 The designer should ensure that the characteristics of the deck water seal, pressure/vacuum breakers and pressure/vacuum valves and the pressure settings of the high and low inert gas deck pressure alarms are compatible. It is also desirable to check that all pressure/vacuum devices are operating at their designed pressure settings.

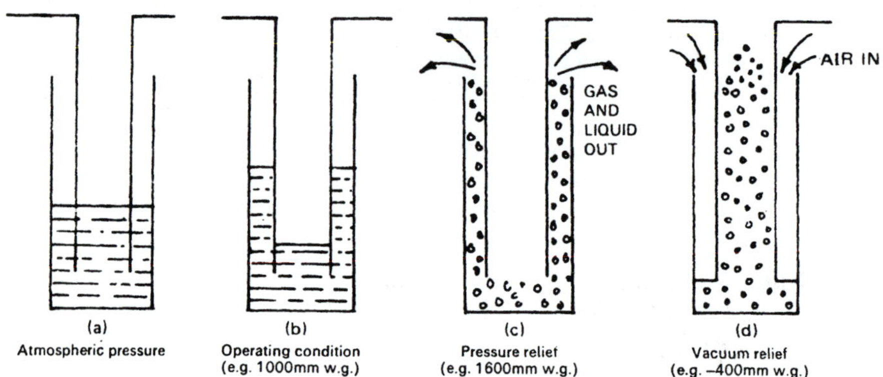

Figure 16 Principles of liquid-filled pressure/vacuum breakers.

3.14 **Instrumentation and alarms** (regulation 62.16 to .19)

3.14.1 Certain fixed and portable instruments are required for the safe and effective operation of an inert gas system. It is desirable that all instruments should be graduated to a consistent system of units.

3.14.2 Clear instructions should be provided for operating, calibrating and testing all instruments and alarms. Suitable calibration facilities should be provided.

3.14.3 All instrumentation and alarm equipment required in compliance with regulation 62 should be suitably designed to withstand supply voltage variation, ambient temperature changes, vibration, humidity, shock, impact and corrosion normally encountered on board ships.

3.14.4 The arrangement of scrubber instrumentation and alarm should be as follows:

.1 The water flow to the scrubber should be monitored either by a flow meter or by pressure gauges. An alarm should be initiated when the water flow drops below the designed flow requirements by a predetermined amount and the inert gas blowers should be stopped automatically in the event of a further reduction in the flow. The precise setting of the alarm and shutdown limits should be related to individual scrubber designs and materials.

.2 The water level in the scrubber shall be monitored by a high water level alarm (see regulation 62.19.1.2). This alarm should be given when predetermined limits are reached and the scrubber pump shut down when the level rises above set limits. These limits should be set having regard to the scrubber design and flooding of the scrubber inlet piping from the boiler uptakes.

.3 The inert gas temperature at the discharge side of the gas blowers shall be monitored. An alarm should be given when the temperature reaches 65°C and automatic shutdown of the inert gas blowers should be arranged if the temperature reaches 75°C.

.4 If a precooler is necessary at the scrubber inlet to protect coating materials in the scrubber, the arrangements for giving an alarm in .3 above should apply to the outlet temperature from the precooler.

.5 To monitor the scrubber efficiency, it is recommended that the cooler water inlet and outlet temperatures, and the scrubber differential pressures are indicated.

.6 All sensing probes, floats and sensors required to be in contact with the water and gas in the scrubber should be made from materials resistant to acidic attack.

3.14.5 For the deck water seal, an alarm should be given when the water level falls by a pre-determined amount but before the seal is rendered ineffective. For certain types of deck water seals, such as the dry type, the water level alarm may require to be suppressed when inert gas is being supplied to the inert gas distribution system.

3.14.6 The pressure of the inert gas in the inert gas main shall be monitored (see regulation 62.19.1.9). An alarm should be given when the pressure reaches the set limit. The set limit should be set having regard to the design of cargo tanks, mechanical nonreturn valve and deck water seal.

3.14.7 The arrangement for oxygen analyser, recorder and indicating equipment should be as follows:

.1 The sampling point for the oxygen analyser and recorder unit should be located at a position in the pipework after the blowers and before the gas pressure regulating valve specified in regulation 62. At the chosen position turbulent flow conditions should prevail at all outputs of the blowers. The sample point should be easily accessible and be provided with suitable air or steam cleaning connections.

.2 The sampling probe should incorporate a dust filter in accordance with the instrument manufacturer's advice. The probe and filter should be capable of being withdrawn and cleaned or renewed as necessary.

.3 The sensing pipe from the sampling probe to the oxygen analyser should be so arranged that any condensation in the sensing pipeline does not prevent the gas sample reaching the oxygen analyser. Joints in the pipeline should be kept to a minimun to prevent the ingress of air.

.4 Any coolers required in the sensing pipes should be installed at the coldest point in the system; alternatively, in certain instances it may be prudent to heat the sensing pipes to prevent condensation.

.5 The position of the analyser should be so chosen that it is protected from heat and adverse ambient conditions, but it should be placed as close as practicable to the sampling point to reduce the time between the extraction of a sample and its analysis to a minimum.

.6 The recording unit and repeater indication required by regulation 62.16 should not be located in positions subject to heat and undue vibration.

.7 The resistance of the connecting cables between the analyser and the recorder should be in accordance with the instrument manufacturer's instruction.

.8 The oxygen analyser should have an accuracy of ±1% of the full-scale deflection of the indicator.

.9 Dependent on the principle of measurement, fixed zero and/or span calibration arrangements should be provided in the vicinity of the oxygen analyser fitted with suitable connections for portable analysers.

3.14.8 A sampling point should be provided between the automatic gas pressure regulating valve and the deck water seal for use with portable instruments.

3.14.9 The inert gas pressure sensor and recorder should obtain the signal from a point in the inert gas main between the deck isolating/nonreturn valve and the cargo tanks (see regulation 62.16.1).

3.14.10 When the pressure in the inert gas main forward of the nonreturn devices falls below 50 mm water gauge means shall be provided to give an audible alarm or to shut down the main cargo pumps automatically (see regulation 62.19.8).

3.14.11 The alarms required by regulation 62.19.1.7 should be given on the navigating bridge and in the machinery space.

3.14.12 In accordance with regulation 62.17 portable instruments shall be provided for measuring oxygen and flammable vapour concentration. With regard to the hydrocarbon vapour meter, it should be borne in mind that meters working on the catalytic filament principle are unsuitable for measuring hydrocarbon concentration in oxygen-deficient atmospheres. Furthermore, meters using this principle cannot measure concentrations of hydrocarbon vapours above the lower flammable limit. It is, therefore, advisable to use meters using a principle which is not affected by oxygen deficiency and which are capable of measuring hydrocarbon concentration in and above the flammable range. For measuring below the lower flammable limit, provided sufficient oxygen is present, the catalytic filament meter is suitable.

3.14.13 All metal parts of portable instruments and sampling tubes requiring to be introduced into tanks should be securely earthed to the ship structure while the instruments and sampling tubes are being used. These portable instruments should be of an intrinsically safe type.

3.14.14 Sufficient tubing etc. should be provided to enable fully representative sampling of a cargo tank atmosphere to be obtained.

3.14.15 Suitable openings should be provided in cargo tanks to enable fully representative samples to be taken from each tank. Where tanks are subdivided by complete or partial wash bulkheads, additional openings should be provided for each such subdivision.

3.15 Effluent and drain piping

3.15.1 The effluent piping from scrubbers and deck water seal drain pipes, where fitted, should be of corrosion-resistant material, or of carbon steel suitably protected internally against the corrosive nature of the fluid.

3.15.2 The scrubber effluent pipe and deck water seal drain pipe, where fitted, should not be led to a common drain pipe and the deck seal drain should be led clear of the engine-room and any other gas-safe space.

3.15.3 Piping made in glass-reinforced plastics of acceptable manufacture, substantial thickness, pressure tested and adequately supported, may be acceptable for effluent piping from scrubbers under the following conditions:

.1 The effluent lines should, as far as possible, be led through cofferdams or ballast tanks and be in accordance with the load line regulations in force.

.2 Where effluent lines are led through machinery spaces the arrangements should include:

.2.1 a valve fitted to a stub piece at the shell and actuated both from inside and outside the machinery space by pneumatic or hydraulic means led through steel piping. The valve should close automatically in the event of failure of the operating media. The valve should have a position indicator. This valve is to be closed at all times when the plant is not in operation as well as in the event of a fire in the machinery space. Suitable instructions to this effect are to be given to the master;

.2.2 a flap nonreturn valve;

.2.3 a short length of steel pipe, or spool piece, lined internally and fitted between the valve referred to in .1 above and the nonreturn valve referred to in .2 above; this is to be fitted with a 12.5 mm diameter flanged drain branch pipe and valve;

.2.4 a further spool piece fitted inboard of, and adjacent to, the nonreturn valve referred to in .2 above, similarly fitted with a drain. (*Note:* the purpose of this arrangement is to enable the tightness of the valves and nonreturn valves referred to in .1 and .2 above to be checked and to facilitate the removal of the nonreturn valve for examination and replacement);

.2.5 means should be provided outside the machinery space for stopping the scrubber pump.

A suitable arrangement is illustrated in figure 17.

3.15.4 A water seal in the shape of a 'U' bend at least 2 m in depth should be fitted at least 2 m below the equipment to be drained. Means should be provided for draining the lowest point of the bend. In addition the seal should be adequately vented to a point above the water level in the scrubber or deck water seal.

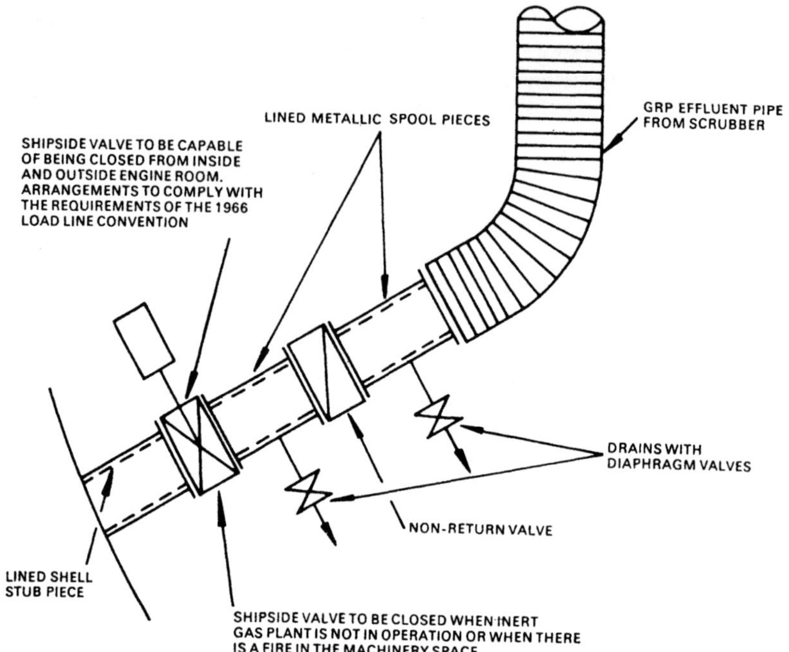

Figure 17 Scrubber seawater supply pumps to be capable of being stopped from outside engine-room.

3.15.5 The diameter of the effluent and drain pipes should be adequate for the duties intended and the pipe run should be self draining from the water seal referred to in 3.15.4.

3.16 Seawater service

3.16.1 It is advisable that the main supply of water to the inert gas scrubber as required in regulation 62.6.1 should be from an independent pump. The alternative source of supply of water may be from another pump such as the sanitary, fire, bilge and ballast pumps provided that the quantity of water required by the inert gas scrubber is readily available and the requirements of other essential services are not thereby impaired.

3.16.2 The requirement for two separate pumps to be capable of supplying water to the deck water seal (see regulation 62.10.3) can be met by any of the pumps referred to under alternative source of supply in 3.16.1 subject to the same provisions applying as are recorded in that paragraph.

3.16.3 The pumps supplying water to the scrubber and the deck water seal should be such as to provide the required throughput of water under light draught conditions. The quantity of water under all other draught conditions should not flood the scrubber or increase the gas flow resistance excessively.

3.16.4 Loops in the piping of the deck water seal to prevent the backflow of hydrocarbon vapour or inert gas should be positioned outside the machinery space and suitably protected against freezing, for example by steam tracing. With reference to the deck water seal arrangement, provisions should be made to prevent a pneumatically controlled system from freezing (see regulation 62.10.5 and .6.)

3.16.5 Vacuum breakers provided to prevent the water loops being emptied should vent to a position on the open deck.

4 OPERATION OF INERT GAS PLANT

4.1 Though flue gas systems differ in detail certain basic principles remain the same. These are:

.1 starting up the inert gas plant;

.2 shutting down the inert gas plant;

.3 safety checks when the inert gas plant is shut down.

In all cases the manufacturer's detailed instructions should be followed.

4.2 Start-up procedures should be as follows:

.1 Ensure boiler is producing flue gas with an oxygen content of 5% by volume or less (for existing ships 8% by volume or, wherever practicable, less).

.2 Ensure that power is available for all control, alarm and automatic shutdown operations.

.3 Ensure that the quantity of water required by the scrubber and deck water seal is being maintained satisfactorily by the pumps selected for this duty.

.4 Test operation of the alarm and shutdown features of the system dependent upon the throughput of water in the scrubber and deck seal.

.5 Check that the gas-freeing fresh air inlet valves, where fitted, are shut and the blanks in position are secure.

.6 Shut off the air to any air sealing arrangements for the flue gas isolating valve.

.7 Open the flue gas isolating valve.

.8 Open the selected blower suction valve. Ensure that the other blower suction and discharge valves are shut unless it is intended to use both blowers simultaneously.

.9 Start the blower.

.10 Test blower "failure" alarm.

.11 Open the blower discharge valve.

.12 Open the recirculating valve to enable plant to stabilize.

.13 Open the flue gas regulating valve.

.14 Check that oxygen content is 5% by volume or less, (for existing ships 8% by volume or, wherever practicable, less) then close the vent to atmosphere between the gas pressure regulating valve and the deck isolating valve.
Note: Some oxygen analysers require as much as two hours to stabilize before accurate readings can be obtained.

.15 The inert gas system is now ready to deliver gas to the cargo tanks.

4.3 Shutdown procedures should be as follows:

.1 When all tank atmospheres have been checked for an oxygen level of not more than 8% and the required in-tank pressure has been obtained, shut the deck isolating/nonreturn valve.

.2 Open vent to atmosphere between the gas pressure regulating valve and the deck isolating/nonreturn valve.

.3 Shut the gas pressure regulating valve.

.4 Shut down the inert gas blower.

.5 Close the blower suction and discharge valve. Check that the drains are clear. Open the water washing system on the blower while it is still rotating with the power supply of the driving motor turned off, unless otherwise recommended by the manufacturer. Shut down the water washing plant after a suitable period.

.6 Close the flue gas isolating valve and open the air sealing system.

.7 Keep the full water supply on the scrubber tower in accordance with the manufacturer's recommendation.

.8 Ensure that the water supply to the deck water seal is running satisfactorily, that an adequate water seal is retained and that the alarm arrangements for it are in order.

4.4 Safety checks when the inert gas plant is shut down should be as follows:

.1 The water supply and water level in the deck seal should be ascertained at regular intervals, at least once per day depending on weather conditions.

.2 Check the water level in water loops installed in pipework for gas, water or pressure transducers, to prevent the backflow of hydrocarbon gases into gas-safe spaces.

.3 In cold weather, ensure that the arrangements to prevent the freezing of sealing water in the deck seals, pressure/vacuum breakers etc. are in order.

.4 Before the pressure in the inerted cargo tanks drops to 100 mm they should be re-pressurized with inert gas.

4.5 Possible failures of the inert gas system and actions to be taken include:

.1 High oxygen content which may be caused or indicated by the following conditions:

.1.1 poor combustion control at the boiler, especially under low load conditions;

.1.2 drawing air down the uptake when boiler gas output is less than the inert gas blower demand, especially under low load conditions;

.1.3 air leaks between the inert gas blower and the boiler uptake;

.1.4 faulty operation or calibration of the oxygen analyser;

.1.5 inert gas plant operating in the recirculation mode; or

.1.6 entry of air into the inert gas main through the pressure vacuum valves, mast risers etc. due to maloperation.

.2 If the inert gas plant is delivering inert gas with an oxygen content of more than 5%, the fault should be traced and repaired. Regulation 62.19.5 requires, however, that all cargo tank operations shall be suspended if the oxygen content exceeds 8% unless the quality of the gas is improved.

.3 Inability to maintain positive pressure during cargo discharge or deballasting operations which may be caused by:

.3.1 inadvertent closure of the inert gas valves;

.3.2 faulty operation of the automatic pressure control system;

.3.3 inadequate blower pressure; or

.3.4 a cargo discharge rate in excess of the blower output.

.4 The cargo discharge or deballasting should be stopped or reduced depending on whether or not the positive pressure in the tanks can be maintained while the fault is rectified.

5 APPLICATION TO CARGO TANK OPERATION

The inert gas system should be used during the full cycle of tanker operation as described in this section.

5.1 Inerting of tanks

5.1.1 Tanks that have been cleaned and gas-freed should be re-inerted preferably during the ballast voyage to allow the inert gas system to be fully tested prior to cargo handling. Purge pipes/vents should be opened to atmosphere. When the oxygen concentration of the atmosphere in the tank has fallen below 8% the purge pipes/vents should be closed and the tank pressurized with inert gas.

5.1.2 During the re-inerting of a tank following a breakdown and repair of the inert gas system, non-gas-free and non-inerted tanks should be re-inerted in accordance with 5.1.1. During inerting, no ullaging, dipping, sampling or other equipment should be inserted unless it has been established that the tank is inert. This should be done by monitoring the efflux gas from the tank being inerted until the oxygen content is less than 8% by volume and for such a period of time as determined by previous test records when inerting gas-free tanks to ensure that the efflux gas is fully representative of the atmosphere within the tank.

5.1.3 When all tanks have been inerted, they should be kept common with the inert gas main and maintained at a positive pressure in excess of 100 mm water gauge during the rest of the cycle of operation.

5.2 Discharge of water ballast

5.2.1 Before discharge of cargo tank ballast is undertaken, the following conditions should be checked:

.1 All cargo tanks are connected up to the inert gas system and all isolating valves in the deck pipework are locked open.

.2 All other cargo tank openings are shut.

.3 All valves isolating the mast risers from the inert gas system are shut.

.4 The arrangements required by regulation 62.13.4.1 are used to isolate the cargo main from the inert gas main.

.5 The inert gas plant is producing gas of an acceptable quality.

.6 The deck isolating valve is open.

5.2.2 During the deballasting operation, the oxygen content of the gas and its pressure in the inert gas main should be continuously recorded (see regulation 62.16.1.1 and .2).

5.3 Loading

When loading cargo the deck isolating valve required by regulation 62.10.8 should be closed and the inert gas plant may be shut down unless other cargo tanks are being deballasted simultaneously. All openings to the cargo tanks except the connections to the mast risers or equivalent venting arrangements should be kept closed to minimize flammable vapour on deck. Before loading commences, the flame screens in the mast risers or equivalent venting arrangements should be inspected and any stop valves isolating the cargo tanks from the inert gas main locked in the open position.

5.4 Loaded condition

5.4.1 During the loaded passage a positive pressure of inert gas of at least 100 mm water gauge should be maintained in the cargo tanks and topping up of the pressure may be necessary. When topping up the inert gas pressure in the cargo tanks, particular attention should be paid to obtaining an oxygen concentration of 5% or less in the inert gas supply before introducing the gas into the cargo tanks.

5.4.2 On motor tankers, the boiler loading may have to be increased in order that the low oxygen concentration in the inert gas supply can be achieved. It may also be necessary to restrict the output of the inert gas blowers to prevent air being drawn down the uptake during the topping up operation. If by these means inert gas of the quality defined in 5.4.1 cannot be achieved then inert gas from an alternative source of supply such as an inert gas generator might be used.

5.5 Cargo transfer and cargo sampling

5.5.1 Ullaging devices of the closed type should be used to avoid the opening of ullage ports.

5.5.2 However, it may be necessary to infrequently relieve the inert gas pressure in the cargo tanks on certain occasions to permit manual tank gauge or cargo sampling before or after cargo is transferred. If this is done, no cargo or ballasting operation is to be undertaken and a minimum number of small tank openings are to be uncovered for as short a time as necessary to enable these measurements to be completed. Manual gauging or cargo sampling may be performed during the following four periods:

.1 At the loading port, prior to cargo loading.

.2 At the loading port, after cargo loading.

.3 At the discharge port, prior to cargo discharge.

.4 At the discharge port, after cargo discharge.

5.5.3 The tanks should then be re-pressurized immediately after the measurements or cargo samples have been taken.

5.5.4 If the tank is opened prior to cargo transfer, cargo transfer should not be commenced until all the conditions have been checked and are in order. Similarly, if the tank is opened after cargo transfer, normal ship operations should not be commenced until all the conditions have been checked and are in order.

5.5.5 During cargo transfer the oxygen content and pressure of the inert gas in the inert gas main should be continuously recorded (see regulation 62.16.1.1 and .2).

5.6 Crude oil washing (see section 5 of the *Crude Oil Washing Operations and Equipment Manual)*

5.6.1 Before each tank is crude oil washed, the oxygen level shall be determined at a point 1 m below the deck and at the middle region of the ullage space and neither of these determinations shall exceed 8% by volume. Where tanks have complete or partial wash bulkhead, the determination should be taken from similar levels in each section of the tank. The oxygen content and pressure of the inert gas being delivered during the washing process should be continuously recorded (see regulation 62.16.1.1 and .2).

5.6.2 If, during the crude oil washing:

.1 the oxygen level of the inert gas being delivered exceeds 8% by volume; or

.2 the pressure of the atmosphere in the tanks is no longer positive;

then washing must be stopped until satisfactory conditions are restored. Operators should also be guided by 4.5.2.

5.7 Ballasting of cargo tanks

The conditions for ballasting of cargo tanks are the same as those for loading in 5.3. When, however, simultaneous discharge and ballasting is adopted, then a close watch should be kept on the inert gas main pressure.

5.8 Ballast condition

5.8.1 During a ballast voyage, tanks other than those required to be gas-free for necessary tank entry should be kept inerted with the cargo tank atmosphere at a positive pressure of not less than 100 mm water gauge having an oxygen level not exceeding 8% by volume especially during tank cleaning.

5.8.2 Before any inert gas is introduced into cargo tanks to maintain a positive pressure it should be established that the inert gas contains not more than 5% by volume of oxygen.

5.9 Tank cleaning

Cargo tanks should be washed in the inert condition and under a positive pressure. The procedures adopted for tank cleaning with water should follow those for crude oil washing in 5.6.

5.10 Purging prior to gas-freeing

When it is desired to gas-free a tank after washing, the concentration of hydrocarbon vapour should be reduced by purging the inerted cargo tank with inert gas. Purge pipes/vents should be opened to atmosphere and inert gas introduced into the tank until the hydrocarbon vapour concentration measured in the efflux gas has been reduced to 2% by volume and until such time as determined by previous tests on cargo tanks has elapsed to ensure that readings have stabilized and the efflux gas is representative of the atmosphere within the tank.

5.11 Gas-freeing

5.11.1 Gas-freeing of cargo tanks should only be carried out when tank entry is necessary (e.g. for essential repairs). It should not be started until it is established that a flammable atmosphere in the tank will not be created as a result. Hydrocarbon gases should be purged from the tank (see 5.10).

5.11.2 Gas-freeing may be effected by pneumatically, hydraulically or steam-driven portable blowers, or by fixed equipment. In either case it is necessary to isolate the appropriate tanks to avoid contamination from the inert gas main.

5.11.3 Gas-freeing should continue until the entire tank has an oxygen content of 21% by volume and a reading of less than 1% of lower flammable limit is obtained on a combustible gas indicator. Care must be taken to prevent the leakage of air into inerted tanks, or of inert gas into tanks which are being gas-freed.

5.12 Tank entry

5.12.1 The entry of personnel to the cargo tank should be carried out only under the close supervision of a responsible ship's officer and in accordance with national rules and/or with the normal industrial practice laid down in the *International Safety Guide for Oil Tankers and Terminals.** The particular hazards encountered in tanks which have been previously inerted and then gas-freed are outlined in 9.2.8, 9.3.3 and chapter 10 of that Guide.

5.12.2 Practical precautions to meet these hazards include:

.1 securing the inert gas branch line gas valves and/or blanks in position or, if gas-freeing with the inert gas blower, isolating the scrubber from the flue gases;

.2 closing of any drain lines entering the tank from the inert gas main;

.3 securing relevant cargo line valves or controls in the closed position;

---

* *ISGOTT* published by the International Chamber of Shipping and the Oil Companies International Marine Forum.

.4 keeping the inert gas deck pressure in the remainder of the cargo tank system at a low positive pressure such as 200 mm water gauge. This minimizes the possible leakage of inert or hydrocarbon gas from other tanks through possible bulkhead cracks, cargo lines, valves, etc.;

.5 lowering clean sample lines well into the lower regions of the tank in at least two locations. These locations should be away from both the inlet and outlet openings used for gas-freeing. After it has been ascertained that a true bottom sample is being obtained, the following readings are required:

.5.1 21% on a portable oxygen analyser; and

.5.2 less than 1% of lower flammable limit on a combustible gas indicator;

.6 the use of breathing apparatus whenever there is any doubt about the tank being gas-free, e.g. in tanks where it is not possible to sample remote locations. (This practice should be continued until all areas, including the bottom structure, have been thoroughly checked);

.7 continuously ventilating and regularly sampling the tank atmosphere whenever personnel are in the tank;

.8 carefully observing normal regulations for tank entry.

5.13 Re-inerting after tank entry

5.13.1 When all personnel have left the tank and the equipment has been removed, the inert gas branch line blank, if fitted, should be removed, the hatch lids closed and the gas pressure regulating valve re-opened and locked open to the inert gas main where appropriate. This will avoid any risk of structural damage when liquids are subsequently handled.

5.13.2 As soon as a gas-free tank is reconnected to the inert gas main it should be re-inerted (as described in 5.1) to prevent transfer of air to other tanks.

6 PRODUCT CARRIERS

The basic principles of inerting are exactly the same on a product carrier as on a crude oil tanker. However, there are differences in operation of these vessels as outlined below.

6.1 **Carriage of products having a flashpoint exceeding 60°C** (closed cup test) as determined by an approved flashpoint apparatus.

6.1.1 Regulation 55(a)(i) of Part E, chapter II-2, 1974 SOLAS Convention* as amended by the 1978 Protocol implies, *inter alia*, that regulations 60 and 62 do not

* Regulation 55.1 of chapter II-2 of the 1983 SOLAS amendments.

apply to tankers carrying petroleum products having a flashpoint exceeding 60°C; in other words, product carriers may carry bitumens, lubricating oils, heavy fuel oils, high flashpoint jet fuels and some diesel fuels, gas oils and special boiling point liquids without inert gas systems having to be fitted, or, if fitted, without tanks containing such cargoes having to be kept in the inert condition.

6.1.2 If cargoes with a flashpoint exceeding 60°C, whether heated or otherwise, are carried at temperatures near to or above their flashpoint (some bitumen cut-backs and fuel oils), a flammable atmosphere can occur (regulation 62.1 refers). When cargoes with a flashpoint exceeding 60°C are carried at a temperature higher than 5°C below their flashpoint they should be carried in an inerted condition.

6.1.3 When a non-volatile cargo is carried in a tank that has not been previously gas-freed, then that tank shall be maintained in an inert condition.

6.2 Product contamination by other cargoes

Contamination of a product may affect its odour, acidity or flashpoint specifications, and may occur in several ways; those relevant to ships with an inert gas main (or other gas line) interconnecting all cargo tanks are:

.1 Liquid contamination due to overfilling a tank.

.2 Vapour contamination through the inert gas main. This is largely a problem of preventing vapour from low flashpoint cargoes, typically gasolines, contaminating the various high flashpoint cargoes listed in 6.1.1, plus aviation gasolines and most hydrocarbon solvents. This problem can be overcome by:

.2.1 removing vapours of low flashpoint cargoes prior to loading; and

.2.2 preventing ingress of vapours of low flashpoint cargoes during loading and during the loaded voyage.

When carrying hydrocarbon solvents where quality specifications are stringent and where it is necessary to keep individual tanks positively isolated from the inert gas main after a cargo has been loaded, pressure sensors should be fitted so that the pressure in each such tank can be monitored. When it is necessary to top up the relevant tanks, the inert gas main should first be purged of cargo vapour.

6.3 Contamination of cargoes by inert gas

For a well designed and operated flue gas system experience suggests that petroleum cargoes traditionally carried on product tankers do not suffer contamination from the flue gas itself, as opposed to contamination from other cargoes. However, unacceptable contamination from the flue gas may be encountered if proper control is not exercised over fuel quality, efficiency of combustion, scrubbing and filtering.

The more critical petrochemical cargoes which may be carried by product carriers can be contaminated by flue gas.

6.4 Contamination of cargoes by water

All lubricating oils and jet fuels are acutely water-critical. Current practice requires full line draining and mopping up of any water in tanks before loading. Water contamination may occur on inerted ships due to:

.1 water carry-over from the scrubber and/or deck water seals due to inadequacies in design or maintenance of the various drying arrangements, and

.2 condensation of water from warm, fully saturated flue gas delivered to the tanks.

6.5 Additional purging and gas-freeing

Gas-freeing is required on non-inerted product carriers more frequently than on crude carriers, both because of the greater need for tank entry and inspection, especially in port, and for venting vapours of previous cargoes. On inerted product carriers any gas-freeing operation has to be preceded by a purging operation (regulation 62.2.4), but gas-freeing for purely quality reasons may be replaced by purging only. In addition purging may be required on the basis outlined in 6.1.3 above.

It should be recognised that:

.1 there are increased risks of air leaking into inert tanks and of inert gas leaking into a tank being entered;

.2 purging is not a prerequisite of gas-freeing when the hydrocarbon gas content of a tank is below 2% by volume;

.3 the operation of gas-freeing for product purity and where tank entry is not contemplated does not require the atmosphere to have an oxygen content of 21% by volume.

7 COMBINATION CARRIERS

The basic principles of inerting are exactly the same on a combination carrier as on a tanker. However, there are differences in the design and operation of these vessels and relevant considerations are outlined below.

7.1 Slack holds

It is particularly important for combination carriers to have their holds inerted because whenever a hold in an OBO carrier (which could extend to the full breadth of the ship) is partially filled with clean or oily ballast, water agitation of this ballast can occur at small angles of roll and this can result in the generation of static electricity. The agitation is sometimes referred to as 'sloshing' and it can happen whenever the ullage above the liquid level of the hold is more than 10% of the depth of the hold, measured from the underside of the deck (see figure 18 for remedy condition).

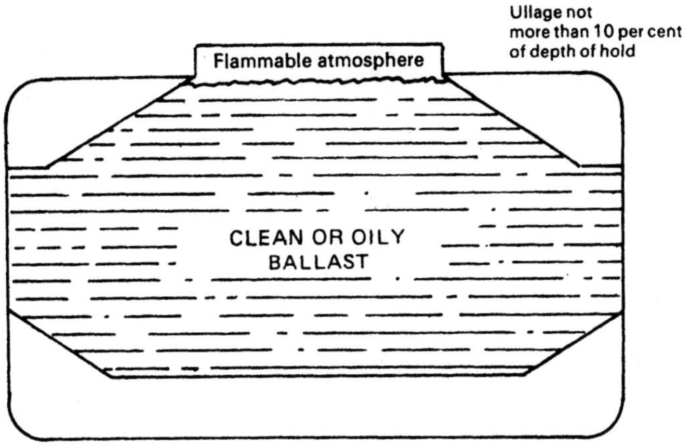

Figure 18

7.2 Leakage

To ensure that leakage of tank gas, particularly through the hatch centre-line joints, is eliminated or minimized, it is essential that the hatch covers are inspected frequently to determine the state of their seals, their alignment, etc. When the hatch covers have been opened, particularly after the ship has been carrying a dry bulk cargo, the seals and trackways should be inspected and cleaned to remove any foreign matter.

7.3 Ballast and void spaces

The cargo holds of combination carriers are adjacent to ballast and void spaces. Leakages may occur in pipelines or ducts in these spaces, or by a fracture in the boundary plating; in this event there is a possibility that oil, inert gas and hydrocarbon gas may leak into the ballast and void spaces. Consequently gas pockets may form and difficulty with gas-freeing should be anticipated due to the considerable steelwork, acting as stiffening, which is characteristic of these spaces. Personnel should be alerted to this hazard.

7.4 Inert gas distribution system

Due to the special construction of combination carriers, the vent line from the cargo hatchway coaming is situated very close to the level of the cargo surface. In many cases, the inert gas main line passing along the main deck may be below the oil level in the hold. During rough weather oil or water may enter these lines and completely block the opening and thus prevent an adequate supply of inert gas during either tank cleaning or discharge. Vent lines should therefore have drains fitted at their lowest point and these should always be checked before any operation takes place within the cargo hold.

7.5 Application when carrying oil

On combination carriers the inert gas system should be utilized in the manner described in section 5 when the ship is engaged exclusively in the carriage of oil.

7.6 Application when carrying cargoes other than oil

7.6.1 When a combination carrier is carrying a cargo other than oil it should be considered as a tanker unless the requirements in 7.6.8 are complied with.

7.6.2 When cargoes other than oil are intended to be carried it is essential that all holds/cargo tanks other than slop tanks referred to in 7.6.6 and 7.6.7 are emptied of oil and oil residues, and cleaned and ventilated to such a degree that the tanks are completely gas-free and internally inspected. The pump-room, cargo pumps, pipelines, duct keel and other void spaces are to be checked to ensure that they are free of oil and hydrocarbon gas.

7.6.3 Where holds are required to carry cargo other than oil they should be isolated from the inert gas main and oil cargo pipeline by means of blanks which should remain in position at all times when cargoes other than oil are being handled or carried.

7.6.4 During the loading and discharging of solid cargoes and throughout the intervening periods all holds/cargo tanks other than the slop tanks referred to in 7.6.6 and 7.6.7, cargo pump-rooms, cofferdams, duct keels and other adjacent void spaces should be kept in a gas-free condition and checked periodically at intervals of not more than two days to ensure that:

- .1 there has been no generation of hydrocarbon gas or leakage of hydrocarbon gas from the slop tanks referred to in 7.6.6 and 7.6.7. If concentrations of more than 20% of the lower flammable limit are detected, the compartments should be ventilated;

- .2 there is no deficiency of oxygen which could be attributable to leakage of inert gas from another compartment.

7.6.5 As an alternative to 7.6.4, those cargo tanks which are empty of cargo may be re-inerted in accordance with 5.1, provided they are subsequently maintained in the inert condition and at a minimum pressure of 100 mm water gauge at all times, and provided that they are checked periodically at intervals of not more than two days to ensure that any generation of hydrocarbon gas does not exceed 1% by volume. If such a concentration is detected the compartments should be purged in accordance with 5.10.

7.6.6 Slops should be contained in a properly constituted slop tank and should be:

- .1 discharged ashore and the slop tanks cleaned and ventilated to such a degree that the tanks are completely gas-free and then inerted; or

- .2 retained on board for not more than one voyage when, unless the vessel reverts to carrying oil, the slop tank should be treated as in 7.6.5.

If slops are retained on board for more than one voyage because reception facilities for oily residues are not available, the slop tank should be treated as in 7.6.5 and in addition a report should be forwarded to the Administration.

7.6.7 Slop tanks which have not been discharged should comply not only with the requirements of 7.6.6, but also with regulation 62.11.2 which requires that they be isolated from other tanks by blank flanges which will remain in position at all times when cargoes other than oil are being carried, except as provided for in these Guidelines. In this connection reference is made to 7.6.3. On combination carriers where there are also empty cargo tanks which are not required to be isolated from the inert gas main then the arrangement for isolating the slop tanks from these tanks should be such as to:

.1 prevent the passage of hydrocarbon gas from the slop tanks to the empty tanks; and

.2 facilitate monitoring (see regulation 62.16.3.1) of and, if necessary, topping up of the pressure in slop tanks and in any empty cargo tanks if the latter are being kept in the inert condition as referred to in 7.6.5.

A suggested arrangement is shown in figure 19.

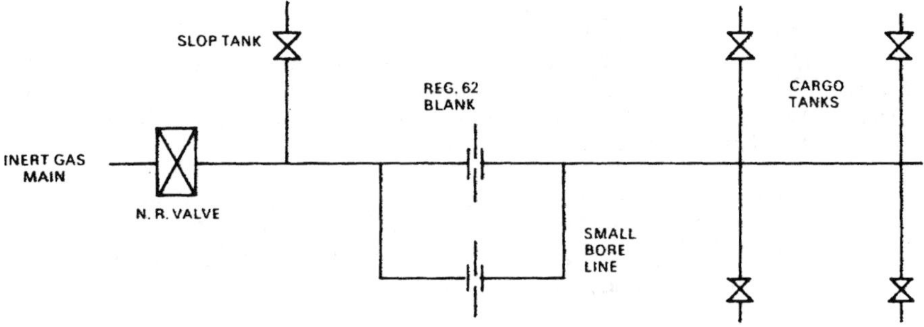

Figure 19 Proposed bypass arrangement for topping up cargo tanks.

In addition, all cargo pipelines to or from the slop tanks should be blanked off.

7.6.8 Instead of complying with the requirements in 7.6.2 to 7.6.7 a combination carrier may be operated as a bulk carrier without having to use its inert gas system if either:

.1 it has never carried a cargo of oil; or

.2 after its last cargo of oil, all its cargo tanks, including slop tanks, the pump-room, cargo pumps, pipelines, cofferdams, duct keel and other void spaces are emptied of oil and oil residues, cleaned and completely gas-free and the tanks and void spaces internally inspected to that effect. In addition the monitoring required in 7.6.4 should be continued until it has been established that generation of hydrocarbon gas has ceased.

8 EMERGENCY PROCEDURES

8.1 In the event of total failure of the inert gas system to deliver the required quality and quantity of inert gas and maintain a positive pressure in the cargo tanks and slop tanks, action must be taken immediately to prevent any air being drawn into the tank. All cargo tank operations should be stopped, the deck isolating valve should be closed, and the vent between it and the gas pressure regulating valve should be opened and immediate action should be taken to repair the inert gas system.

8.2 In the case of tankers engaged in the carriage of crude oil it is essential that the cargo tanks be maintained in the inerted condition to avoid the hazard of pyrophoric iron sulphide ignition. If it is assessed that the tanks cannot be maintained in an inerted condition before the inert gas system can be repaired, an external supply of inert gas should be connected to the system through the arrangements required by regulation 62.11.5 as soon as practicable, to avoid air being drawn into the cargo tanks.

8.3 In the case of product carriers, if it is considered to be totally impracticable to effect a repair to enable the inert gas system to deliver the required quality and quantity of gas and maintain a positive pressure in the cargo tanks, cargo discharge and deballasting may only be resumed provided that either an external supply of inert gas is connected to the system through the arrangements required by regulation 62.11.5, or the following precautions are taken:

.1 In the case of tankers built on or after 1 September 1984, the venting system is checked to ensure that approved devices to prevent the passage of flame into cargo tanks are fitted and that these devices are in a satisfactory condition.

.2 In the case of tankers built before 1 September 1984 the flame screens are checked to ensure that they are in a satisfactory condition.

.3 The valves on the vent mast risers are opened.

.4 No free fall of water or slops is permitted.

.5 No dipping, ullaging, sampling or other equipment should be introduced into the tank unless essential for the safety of the operation. If it is necessary for such equipment to be introduced into the tank, this should be done only after at least 30 minutes have elapsed since the injection of inert gas ceased. All metal components of equipment to be introduced into the tank should be securely earthed. This restriction should be applied until a period of five hours has elapsed since injection of inert gas has ceased.

8.4 In the case of product carriers, if it is essential to clean tanks following a failure of the inert gas system and inerted conditions as defined in regulation 62.2.2 cannot be maintained, tank cleaning should be carried out with an external supply of inert gas connected to the system. Alternatively, if an external supply of inert gas is not connected to the ship, the following precautions should be taken, in addition to 8.3.1 to 8.3.5:

.1 Tank washing should be carried out only on one tank at a time.

.2 The tank should be isolated from other tanks and from any common venting system or the inert gas main and maximum ventilation output should be concentrated on that tank both before and during the washing process. Ventilation should provide as far as possible a free flow of air from one end of the tank to the other.

.3 The tank bottom should be flushed with water and stripped. The piping system including cargo pumps, cross-overs and discharge lines should also be flushed with water.

.4 Washing should not commence until tests have been made at various levels to establish that the vapour content in any part of the tank is below 10% of the lower flammable limit.

.5 Testing of the tank atmosphere should continue during the washing process. If the vapour level rises to within 50% of the lower flammable limit, washing should be discontinued until the vapour level has fallen to 20% of the lower flammable limit or less.

.6 If washing machines with individual capacities exceeding 60 m^3/h are to be used, only one such machine shall be used at any one time on the ship. If portable machines are used, all hose connections should be made up and bonding cables tested for continuity before the machines are introduced into the tank and should not be broken until after the machines have been removed from the tank.

.7 The tank should be kept drained during washing. If build-up of wash water occurs, washing should be stopped until the water has been cleared.

.8 Only clean, cold seawater should be used. Recirculating systems should not be used.

.9 Chemical additives should not be used.

.10 All deck openings, except those necessary for washing and designed venting arrangements, should be kept closed during the washing process.

8.5 During cargo operations in port, more stringent regulations of the port Authorities shall take precedence over any of the foregoing emergency procedures.

8.6 The attention of the ship's master should be drawn to regulation 11(c) of chapter 1 of the 1978 SOLAS Protocol in the event of the inert gas system having become inoperative.

9 MAINTENANCE AND TESTING

9.1 General

9.1.1 The safety arrangements are an integral part of the inert gas system and it is important for ship's staff to give special attention to them during any inspection.

9.1.2 Inspection routines for some of the main components are dealt with in this section.

9.2 Inert gas scrubber

9.2.1 Inspection may be made through the manholes. Checks should be made for corrosion attacks, fouling and damage to:

- .1 scrubber shell and bottom;
- .2 cooling water pipes and spray nozzles (fouling);
- .3 float switches and temperature sensors;
- .4 other internals such as trays and demister filters.

9.2.2 Checks should be made for damage to non-metallic parts such as:

- .1 internal linings;
- .2 demisters;
- .3 packed beds.

9.3 Inert gas blowers

9.3.1 To a limited degree, internal visual inspection will reveal damage at an early stage. Diagnostic monitoring systems should be used as they greatly assist in maintaining the effectiveness of the equipment. By fitting two equal-sized blowers or, alternatively, supplying and retaining on board a spare impeller with a shaft for each blower, an acceptable level of availability is ensured. Visual inspection through the available openings in the blower casing is adequate for this purpose.

9.3.2 An inspection of inert gas blowers should include:

- .1 internal inspection of the blower casing for soot deposits or signs of corrosive attack;
- .2 examination of fixed or portable washing system;
- .3 inspection of the functioning of the fresh water flushing arrangements, where fitted;
- .4 inspection of the drain lines from the blower casing to ensure that they are clear and operative;
- .5 observation of the blower under running conditions for signs of excessive vibration, indicating too large an imbalance.

9.4 Deck water seal

9.4.1 This unit performs an important function and must be maintained in good condition. Corroded inlet pipes and damage to float-controlled valves are not uncommon. The overboard drain line and connection are also possible sources of trouble.

9.4.2 An inspection of the deck water seal should include:

.1 Opening for internal inspection to check for:

.1.1 blockage of the venturi lines in semi-dry type water seals;

.1.2 corrosion of inlet pipes and housing;

.1.3 corrosion of heating coils;

.1.4 corroded or sticking floats for water drain and supply valves and level monitoring.

.2 Testing for function:

.2.1 automatic filling and draining: check with a local level gauge if possible;

.2.2 presence of water carry-over (open drain cocks on inert gas main line) during operation.

9.5 Nonreturn valve

The nonreturn valve should be opened for inspection to check for corrosion and also to check the condition of the valve seat. The functioning of the valve should be tested in operation.

9.6 Scrubber effluent line

The scrubber effluent line cannot normally be inspected internally except when the ship is in dry dock. The shipside stub piece, referred to in 3.15.3.2 and the overboard discharge valve should be inspected at each dry-docking period.

9.7 Testing of other units and alarms

9.7.1 A method should be devised to test the correct functioning of all units and alarms and it may be necessary to simulate certain conditions to carry out an effective testing programme.

9.7.2 Such a programme should include checking:

.1 all alarm and safety functions;

.2 the functioning of the flue gas isolating valves;

.3 the operation of all remotely or automatically controlled valves;

.4 the functioning of the water seal and nonreturn valve (with a backflow pressure test);

.5 the vibration level of the inert gas blowers;

.6 for leakages: in systems four years old or more, deck lines should be examined for gas leakage;

.7 the interlocking of the soot blowers;

.8 oxygen measuring equipment, both portable and fixed, for accuracy by means of both air and a suitable calibration gas.

9.8 Suggested maintenance programme

Component	Preventive maintenance	Maintenance interval
Flue gas isolating valves	Operate the valve	Before start-up and one week
	Cleaning with compressed air or steam	Before operating valve
	Dismantling for inspection and cleaning	Boiler shutdown
Flue gas scrubber	Water flush	After use
	Cleaning of demister	Three months
	Dismantling of level regulators and temperature probes for inspection	Six months
	Opening for full internal inspection	Dry-docking
Overboard pipes and valve from flue gas scrubber	Flushing with scrubber water pump for about one hour	After use
	Dismantling of the valve for overhaul, inspection of pipeline and overboard end	Dry-docking/repair period
Blowers	Vibration check	While running
	Flushing	After use
	Internal inspection through hatches	After flushing and six months

Component	Preventive maintenance	Maintenance interval
Blowers *(cont.)*	Dismantling for full overhaul of bearings, shaft tightenings and other necessary work	Two years or more frequently if required/dry-docking
Deck water seal	Dismantling of level regulators/float valves for inspection	Six months
	Opening for total internal inspection	One year
	Overhaul of auto-valves	One year
Deck mechanical nonreturn valve	Moving and lubricating the valve is necessary	One week and before start
	Opening for internal inspection	One year/18 months
Pressure/ vacuum valves	Operating and lubricating the valves	Six months
	Opening for full overhaul and inspection	One year
Deck isolating valve	Opening for overhaul	One year
Gas pressure regulating system	Removal of condensation in instrument, air supply	Before start
	Opening of gas pressure regulating valves for overhaul	As appropriate
Liquid-filled pressure/ vacuum breaker	Check liquid level when system is at atmospheric pressure	When opportunity permits and every six months

10 TRAINING

10.1 General

10.1.1 An inert gas installation is an important feature of a tanker's safety system and training in its use is essential.

10.1.2 The requirements for training depend upon the policies of the shipping company concerned as well as the Administration of the country in which the ship is registered. This chapter is not intended to specify any particular training policy but to set out a number of alternatives which can be suitably adapted.

10.2 Personnel requiring training

10.2.1 It is not the intention of this section to spell out in detail a syllabus for courses in the design, operation and maintenance of inert gas systems, but it is suggested that any syllabus should cover the same ground as that contained in these Guidelines.

10.2.2 However, such practical training can be given only if those in charge of, and responsible for, the vessel's safety and performance are themselves completely familiar with the type of installation fitted, and the hazards associated with use. It is recommended that the training of both deck and engine-room personnel be co-ordinated to ensure a common understanding of the procedures.

10.2.3 Administrations should make sure that the vessel is equipped with the necessary manufacturers' manuals and instructions to give the necessary information about how to carry out the various operations.

10.3 Location of training

Training may take place aboard or ashore. If shore training in basic design and operation is given, personnel should be made familiar with the equipment on board ship.

10.4 Some training methods

There are currently three methods used in training. Companies may practise one, or a permutation of the following:

.1 *On board by shipping company staff*

This may be carried out either by a senior member of the ship's company who has been made responsible for training or by a specialist who joins the vessel for part of a voyage. Such a training programme can be enhanced by films and other suitable audio-visual aids, if they are available. Under these circumstances, it should be possible to deal with the theoretical as well as the practical aspects.

.2 *Specialist shore-based training*

This can be undertaken by nautical colleges either in consultation with shipping companies or with manufacturers.

It has been found that a one-week course should cover the subject adequately.

.3 *Shore-based by shipping company staff*

Training under this heading may occur either as part of a company cargo course, or, for example, as part of a senior officer's seminar where appropriate time may be devoted to a discussion of inert gas and operating problems.

11 INSTRUCTION MANUAL(S)

Instruction manuals required to be provided on board by regulation 62(u) should contain the following information and operational instructions.

11.1 A line drawing of the inert gas system showing the positions of the inert gas pipework from the boiler or gas generator uptakes to each cargo tank and slop tank; gas scrubber; scrubber cooling water pump and pipework up to the effluent discharge overboard; blowers including the suction and discharge valves; recirculation or other arrangements to stabilize the inert gas plant operation; fresh air inlets; automatic gas pressure regulating stop valve; deck water seal and water supply, heating and overflow arrangements; deck nonreturn stop valve; water traps in any supply, vent, drain and sensing pipework; cargo tank isolation arrangement; purge pipes/vents; pressure/vacuum valves on tanks; pressure/vacuum breakers on the inert gas main; permanent recorders and instruments and the take-off points for their use, arrangements for using portable instruments, complete and partial wash bulkheads, mast risers, mast riser isolating valves; high velocity vents; manual and remote controls.

11.2 A description of the system and a listing of procedures for checking that each item of the equipment is working properly during the full cycle of the tanker operation. This includes a listing of the parameters to be monitored such as inert gas main pressure, oxygen concentration in the delivery main, oxygen concentration in the cargo tanks, temperature at the scrubber outlet and blower outlet, blower running current or power, scrubber pump running current or power, deck seal level during inert gas discharge to cargo tanks at maximum rate, deck seal level at nil discharge, etc. Established values for these parameters during acceptance trials should be included, where relevant.

11.3 Detailed requirements for conducting the operations described in sections 4 and 5 particular to the installation of the ship such as times to inert, purge and gas-free each tank, sequence and number of tanks to be inerted, purged and gas-freed, sequence and number of purge pipes/vents to be opened or closed during such operations, etc.

11.4 Dangers of leakage of inert gas and hydrocarbon vapours and precautions to be taken to prevent such leakages should be described relating to the particular construction and equipment on board.

11.5 Dangers of cargo tank overpressure and underpressure during the various stages in the cycle of tanker operation and the precautions to be taken to prevent such conditions from arising should also be described in detail relating to the particular construction or the equipment on board.

12 SOME SAFETY CONSIDERATIONS WITH INERT GAS SYSTEMS

12.1 Backflow of cargo gases

To prevent the return of cargo gases or cargo from the tanks to the machinery spaces and boiler uptake, it is essential that an effective barrier is always present between these two areas. In addition to a nonreturn valve, a water seal and vent should be fitted on the deck main. It is of prime importance that these devices are properly maintained and correctly operated at all times.

An additional water seal is sometimes fitted at the bottom of the scrubber (see also 3.9.6).

12.2 Health hazards

12.2.1 *Oxygen deficiency*

Exposure to an atmosphere with a low concentration of oxygen does not necessarily produce any recognizable symptom before unconsciousness occurs, when the onset of brain damage and risk of death can follow within a few minutes. If the oxygen deficiency is not sufficient to cause unconsciousness, the mind is liable to become apathetic and complacent, and even if these symptoms are noticed and escape is attempted, physical exertion will aggravate the weakness of both mind and body. It is therefore necessary to ventilate thoroughly to ensure that no pockets of oxygen-deficient atmosphere remain. When testing for entry a steady reading of 21% oxygen is required.

12.2.2 *Toxicity of hydrocarbon vapours*

Inert gas does not affect the toxicity of hydrocarbon gases and the problem of toxicity is no different from that of ships without an inert gas system. Because of possible gas pockets, regeneration, etc. gas-freeing must continue until the entire compartment shows a zero reading with a reliable combustible gas indicator or equivalent, or a 1% of the lower flammable limit reading should the instrument have a sensitivity scale on which a zero reading is not obtainable.

12.2.3 *Toxicity of flue gas*

The presence of toxic gases such as sulphur dioxide, carbon monoxide of nitrogen, can be ascertained only by measurement. However, provided that the hydrocarbon gas content of an inerted tank exceeds about 2% by volume before gas-freeing is started, the dilution of the toxic components of flue gas during the subsequent gas-freeing can be correlated with the readings of an approved combustible gas indicator or equivalent. If by ventilating the compartment, a reading of 1% of the lower flammable limit or less is obtained in conjunction with an oxygen reading of 21% by volume, the toxic trace gases will be diluted to concentrations at which it will be safe to enter. Alternatively, and irrespective of initial hydrocarbon gas content, ventilation should be continued until a steady oxygen reading of 21% by volume is obtained.

12.3 Tank pressure

When an inerted cargo tank is maintained at a positive pressure, personnel should be advised of the practical hazards of this pressure. Such pressure must be adequately reduced before any tank-lids, ullage plugs or tank washing openings are opened.

12.4 Electrostatic hazards

12.4.1 Small particulate matter carried in flue gas can be electrostatically charged. The level of charge is usually small, but levels have been observed well above those encountered with water mists formed during tank washing.

12.4.2 Because cargo tanks are normally in an inerted condition, the possibility of electrostatic ignition has to be considered only if the oxygen content of the tank atmosphere rises as a result of an ingress of air or if it is necessary to inert a tank which already has a flammable atmosphere (see 5.1).

12.5 Repair of inert gas plant

12.5.1 Inert gas is asphyxiating. Great care must be exercised when work on the plant is undertaken. Although the worker may be in the open air, inert gas leaking from the plant could render him unconscious very quickly. Before opening up any equipment, therefore, it is recommended that the inert gas plant is completely gas-freed.

12.5.2 If any unit (e.g. the inert gas scrubber) is to be examined internally, the standard recommendations for entering enclosed spaces must be followed. Blind flanges should be fitted where applicable or the plant should be completely isolated.

12.6 Hazards from pyrophoric iron sulphide

Bearing in mind the reduction of oxygen in ullage spaces compounded by the operation of inert gas systems, research has led to the conclusion that there is a significant risk of pyrophoric deposits forming in inerted tankers carrying sour crude oil; furthermore, that pyrophoric deposits can form with crude oils having a low hydrogen sulphide content and that no minimum safe level of hydrogen sulphide content can be identified; and, finally, that pyrophors which have formed during a loaded passage can persist during the subsequent ballast voyage.

Thus, while various factors (such as lack of sufficiently thick deposits of iron oxide) may inhibit pyrophor formation and while the correct operation of the inert gas plant will prevent the possibility of ignition, the degree of risk is judged to be sufficiently high to require the precautions in section 8.2 in case of inert gas system failure.

PART II PROVISIONS OF APPLICATION

1. 1974 SOLAS Convention
(Entry into force: 25 May 1980)

CHAPTER II-2

CONSTRUCTION – FIRE PROTECTION, FIRE DETECTION AND FIRE EXTINCTION

PART A – GENERAL*

Regulation 1

Application

(a) For the purpose of this Chapter:

(i) A new passenger ship is a passenger ship the keel of which is laid or which is at a similar stage of construction on or after the date of coming into force of the present Convention, or a cargo ship which is converted to a passenger ship on or after that date, all other passenger ships being considered as existing ships.

(ii) A new cargo ship is a cargo ship the keel of which is laid or which is at a similar stage of construction on or after the date of coming into force of the present Convention.

(iii) A ship which undergoes repairs, alterations, modifications and outfitting related thereto shall continue to comply with at least the requirements previously applicable to the ship. An existing ship in such a case shall not as a rule comply to a lesser extent with the requirements for a new ship than it did before. Repairs, alterations and modifications of a major character and outfitting related thereto should meet the requirements for a new ship in so far as the Administration deems reasonable and practicable.

(b) Unless expressly provided otherwise:

(i) Regulations 4 to 16 of Part A of this chapter apply to new ships.

---

* Reference is made to Recommendation on Safety Measures for Periodically Unattended Machinery Spaces of Cargo Ships additional to those normally considered necessary for an Attended Machinery Space, adopted by the Organization by resolution A.211(VII).

(ii) Part B of this chapter applies to new passenger ships carrying more than 36 passengers.

(iii) Part C of this chapter applies to new passenger ships carrying not more than 36 passengers.

(iv) Part D of this chapter applies to new cargo ships.

(v) Part E of this chapter applies to new tankers.

(c) (i) Part F of this chapter applies to existing passenger ships carrying more than 36 passengers.

(ii) Existing passenger ships carrying not more than 36 passengers and existing cargo ships shall comply with the following:

(1) for ships the keels of which were laid or which were at a similar stage of construction on or after the date of coming into force of the International Convention for the Safety of Life at Sea, 1960, the Administration shall ensure that the requirements which were applied under chapter II of that Convention to new ships as defined in that chapter are complied with;

(2) for ship the keels of which were laid or which were at a similar stage of construction on or after the date of coming into force of the International Convention for the Safety of Life at Sea, 1948, but before the date of coming into force of the International Convention for the Safety of Life at Sea, 1960, the Administration shall ensure that the requirements which were applied under chapter II of the 1948 Convention to new ships as defined in that chapter are complied with;

(3) for ships the keels of which were laid or which were at a similar stage of construction before the date of coming into force of the International Convention for the Safety of Life at Sea, 1948, the Administration shall ensure that the requirements which were applied under chapter II of that Convention to existing ships as defined in that chapter are complied with.

(d) For any existing ship as defined in the present Convention the Administration, in addition to applying the requirements of subparagraph (c)(i) of this regulation, shall decide which of the requirements of this chapter not contained in chapter II of the 1948 and 1960 Conventions shall be applied.

(e) The Administration may, if it considers that the sheltered nature and conditions of the voyage are such as to render the application of any specific requirements of this chapter unreasonable or unnecessary, exempt from those requirements individual ships or classes of ships belonging to its country which, in the course of their voyage, do not proceed more than 20 miles from the nearest land.

(f) In the case of passenger ships which are employed in special trades for the carriage of large numbers of special trade passengers, such as the pilgrim trade, the Administration, if satisfied that it is impracticable to enforce compliance with the requirements of this chapter, may exempt such ships, when they belong to its country, from those requirements, provided that they comply fully with the provisions of:

(i) the Rules annexed to the Special Trade Passenger Ships Agreement, 1971, and

(ii) the Rules annexed to the Protocol on Space Requirements for Special Trade Passenger Ships, 1973, when it comes into force.

PART E – FIRE SAFETY MEASURES FOR TANKERS

Regulation 55

Application

(a) This Part shall apply to all new tankers carrying crude oil and petroleum products having a flashpoint not exceeding 60°C (140°F) (closed cup test) as determined by an approved flashpoint apparatus and whose Reid vapour pressure is below that of atmospheric pressure, and other liquid products having a similar fire hazard.

(b) In addition, all ships covered by this Part shall comply with the requirements of regulations 52, 53 and 54 of this chapter, except that paragraph (f) of regulation 52 need not apply to tankers complying with regulation 60 of this chapter.

(c) Where cargoes other than those referred to in paragraph (a) of this regulation which introduce additional fire hazards are intended to be carried, additional safety measures shall be required to the satisfaction of the Administration.

(d) Combination carriers shall not carry solid cargoes unless all cargo tanks are empty of oil and gas freed or unless, in each case, the Administration is satisfied with the arrangements provided.

Regulation 60

Cargo tank protection

(a) For tankers of 100,000 metric tons deadweight and upwards and combination carriers of 50,000 metric tons deadweight and upwards, the protection of the cargo tanks deck area and cargo tanks shall be achieved by a fixed deck froth system and a fixed inert gas system in accordance with the requirements of regulations 61 and 62 of this Part except that in lieu of the above installations the Administration, after having given consideration to the ship arrangement and equipment, may accept other combinations of fixed installations if they afford protection equivalent to the above, in accordance with regulation 5 of chapter I of this Convention.

(b) To be considered equivalent, the system proposed in lieu of the deck froth system shall:

(i) be capable of extinguishing spill fires and also preclude ignition of spilled oil not yet ignited; and

(ii) be capable of combating fires in ruptured tanks.

(c) To be considered equivalent, the system proposed in lieu of the fixed inert gas system shall:

(i) be capable of preventing dangerous accumulations of explosive mixtures in intact cargo tanks during normal service throughout the ballast voyage and necessary in-tank operations; and

(ii) be so designed as to minimize the risk of ignition from the generation of static electricity by the system itself.

(d) In tankers of less than 100,000 metric tons deadweight and combination carriers of less than 50,000 metric tons deadweight the Administration, in applying the requirements of paragraph (f) of regulation 52 of this chapter, may accept a froth system, capable of discharging froth internally or externally, to the tanks. The details of such installation shall be to the satisfaction of the Administration.

2. *1978 SOLAS Protocol*

(Entry into force: 1 May 1981)

CHAPTER II-2

CONSTRUCTION – FIRE PROTECTION, FIRE DETECTION AND FIRE EXTINCTION

PART A – GENERAL

Regulation 1

Application

The following subparagraphs are added to the existing text of paragraph (a):

(iv) Notwithstanding the provisions of subparagraphs (ii) and (iii) of this paragraph, for the purposes of paragraph (a)(ii) of regulation 55 and of regulation 60 of this chapter, a new tanker means a tanker:

(1) for which the building contract is placed after 1 June 1979; or

(2) in the absence of a building contract, the keel of which is laid, or which is at a similar stage of construction after 1 January 1980; or

(3) the delivery of which is after 1 June 1982; or

(4) which has undergone an alteration or modification of a major character:

(a) for which the contract is placed after 1 June 1979; or

(b) in the absence of a contract, the construction work of which is begun after 1 January 1980; or

(c) which is completed after 1 June 1982.

(v) For the purposes of paragraph (a)(ii) of regulation 55 and of regulation 60 of this chapter, an existing tanker is a tanker which is not a new tanker as defined in subparagraph (iv) of this paragraph.

(vi) For the purposes of subparagraph (iv) of this paragraph, conversion of an existing tanker of 20,000 metric tons deadweight and upwards to meet the requirements of the present Protocol or the Protocol of 1978 Relating to the International Convention for the Prevention of Pollution from Ships, 1973 shall not be deemed to constitute an alteration or modification of a major character.

PART E - FIRE SAFETY MEASURES FOR TANKERS

Regulation 55

Application

The existing text of this regulation is replaced by the following:

(a) Unless expressly provided otherwise:

(i) this Part shall apply to all new tankers carrying crude oil and petroleum products having a flashpoint not exceeding 60°C (140°F) (closed cup test) as determined by an approved flashpoint apparatus and a Reid vapour pressure which is below atmospheric pressure and other liquid products having a similar fire hazard; and

(ii) in addition, all ships covered by this Part shall comply with the requirements of regulations 52, 53 and 54 of chapter II-2 of the Convention except that fixed gas fire-extinguishing systems for cargo spaces shall not be used for new tankers and for those existing tankers complying with regulation 60 of this chapter. For existing tankers not required to comply with regulation 60, the Administration, in applying the requirements of paragraph (f) of regulation 52, may accept a froth system capable of discharging froth internally or externally to the tanks. The details of the installation shall be to the satisfaction of the Administration.

(b) Where cargoes other than those referred to in subparagraph (a)(i) of this regulation which introduce additional fire hazards are intended to be carried, additional safety measures shall be required to the satisfaction of the Administration.

(c) Combination carriers shall not carry solid cargoes unless all cargo tanks are empty of oil and gas freed or unless, in each case, the Administration is satisfied with the arrangements provided.

Regulation 60

Cargo tank protection

The existing text of this regulation is replaced by the following:

(a) For new tankers of 20,000 metric tons deadweight and upwards, the protection of the cargo tanks deck area and cargo tanks shall be achieved by a fixed deck froth system and a fixed inert gas system in accordance with the requirements of regulations 61 and 62 of chapter II-2 of the Convention except that in lieu of the above installations the Administration, after having given consideration to the ship's arrangement and equipment, may accept other combinations of fixed installations if they afford protection equivalent to the above, in accordance with regulation 5 of chapter I of the Convention.

(b) To be considered equivalent, the system proposed in lieu of the deck froth system shall:

(i) be capable of extinguishing spill fires and also preclude ignition of spilled oil not yet ignited; and

(ii) be capable of combating fires in ruptured tanks.

(c) To be considered equivalent, the system proposed in lieu of the fixed inert gas system shall:

(i) be capable of preventing dangerous accumulations of explosive mixtures in intact cargo tanks during normal service throughout the ballast voyage and necessary in-tank operations; and

(ii) be so designed as to minimize the risk of ignition from the generation of static electricity by the system itself.

(d) Any existing tanker of 20,000 metric tons deadweight and upwards engaged in the trade of carrying crude oil shall be fitted with an inert gas system, complying with the requirements of paragraph (a) of this regulation, not later than a date:

(i) for a tanker of 70,000 metric tons deadweight and upwards, two years after the date of entry into force of the present Protocol; and

(ii) for a tanker of less than 70,000 metric tons deadweight, four years after the date of entry into force of the present Protocol, except that for tankers less than 40,000 tons deadweight not fitted with tank washing machines having an individual throughput of greater than 60 cubic metres per hour, the Administration may exempt existing tankers from the requirements of this paragraph, if it would be unreasonable and impracticable to apply these requirements, taking into account the ship's design characteristics.

(e) Any existing tanker of 40,000 metric tons deadweight and upwards engaged in the trade of carrying oil other than crude oil and any such tanker of 20,000 metric tons deadweight and upwards engaged in the tráde of carrying oil other than crude oil fitted with tank washing machines having an individual throughput of greater than 60 cubic metres per hour shall be fitted with an inert gas system, complying with the requirements of paragraph (a) of this regulation, not later than a date:

(i) for a tanker of 70,000 metric tons deadweight and upwards, two years after the date of entry into force of the present Protocol; and

(ii) for a tanker of less than 70,000 metric tons deadweight, four years after the date of entry into force of the present Protocol.

(f) Any tanker operating with a cargo tank cleaning procedure using crude oil washing shall be fitted with an inert gas system complying with the requirements of regulation 62 of chapter II-2 of the Convention and with fixed tank washing machines.

(g) All tankers fitted with a fixed inert gas system shall be provided with a closed ullage system.

(h) Any new tanker of 2,000 tons gross tonnage and upwards not covered by paragraph (a) of this regulation shall be provided with a froth system, capable of discharging froth internally or externally, to the tanks. The details of such installation shall be to the satisfaction of the Administration.

3. *1981 Amendments to the 1974 SOLAS Convention*
(Entry into force: 1 September 1984)

CHAPTER II-2

CONSTRUCTION – FIRE PROTECTION, FIRE DETECTION AND FIRE EXTINCTION

The existing text of chapter II-2 is replaced by the following:

PART A – GENERAL

Regulation 1

Application

1.1 Unless expressly provided otherwise, this chapter shall apply to ships the keels of which are laid or which are at a similar stage of construction on or after 1 September 1984.

1.2 For the purpose of this chapter the term "a similar stage of construction" means the stage at which:

.1 construction identifiable with a specific ship begins; and

.2 assembly of that ship has commenced comprising at least 50 tonnes or 1 per cent of the estimated mass of all structural material, whichever is less.

1.3 For the purpose of this chapter:

.1 the expression "ships constructed" means "ships the keels of which are laid or which are at a similar stage of contruction";

.2 the expression "all ships" means "ships constructed before, on or after 1 September 1984";

.3 a cargo ship, whenever built, which is converted to a passenger ship shall be treated as a passenger ship constructed on the date on which such a conversion commences.

2 Unless expressly provided otherwise:

.1 for ships constructed before 1 September 1984, the Administration shall ensure that, subject to the provisions of paragraph 2.2, the requirements which are applicable under chapter II-2 of the International Convention for the Safety of Life at Sea, 1974* to new or existing ships as defined in that chapter are complied with;

.2 for tankers constructed before 1 September 1984, the Administration shall ensure that the requirements which are applicable under chapter II-2 of the Annex to the Protocol of 1978 relating to the International Convention for the Safety of Life at Sea, 1974, to new or existing ships as defined in that chapter are complied with.

3 All ships which undergo repairs, alterations, modifications and outfitting related thereto shall continue to comply with at least the requirements previously applicable to these ships. Such ships, if constructed before 1 September 1984 shall, as a rule, comply with the requirements for ships constructed on or after that date to at least the same extent as they did before undergoing such repairs, alterations, modifications or outfitting. Repairs, alterations and modifications of a major character and outfitting related thereto shall meet the requirements for ships constructed on or after 1 September 1984 in so far as the Administration deems reasonable and practicable.

4.1 The Administration of a State may, if it considers that the sheltered nature and conditions of the voyage are such as to render the application of any specific requirements of this chapter unreasonable or unnecessary, exempt from those requirements individual ships or classes of ships entitled to fly the flag of that State which, in the course of their voyage, do not proceed more than 20 miles from the nearest land.

4.2 In the case of passenger ships which are employed in special trades for the carriage of large numbers of special trade passengers, such as the pilgrim trade, the Administration of the State whose flag such ships are entitled to fly, if satisfied that it is impracticable to enforce compliance with the requirements of this chapter, may exempt such ships from those requirements, provided that they comply fully with provisions of:

.1 the Rules annexed to the Special Trade Passenger Ships Agreement, 1971; and

.2 the Rules annexed to the Protocol on Space Requirements for Special Trade Passenger Ships, 1973.

* The text as adopted by the International Conference on Safety of Life at Sea, 1974.

PART D - FIRE SAFETY MEASURES FOR TANKERS

(The requirements of this Part are additional to those of Part C except for regulations 53 and 54 which do not apply to tankers and except as provided otherwise in regulations 57 and 58)

Regulation 55

Application

1 Unless expressly provided otherwise, this Part shall apply to tankers carrying crude oil and petroleum products having a flashpoint not exceeding 60°C (closed cup test), as determined by an approved flashpoint apparatus, and a Reid vapour pressure which is below atmospheric pressure and other liquid products having a similar fire hazard.

2 Where liquid cargoes other than those referred to in paragraph 1 or liquefied gases which introduce additional fire hazards are intended to be carried, additional safety measures shall be required to the satisfaction of the Administration, having due regard to the provisions of the Bulk Chemical Code and the Gas Carrier Code.

3 This paragraph applies to all ships which are combination carriers. Such ships shall not carry solid cargoes unless all cargo tanks are empty of oil and gas freed or unless the arrangements provided in each case are to the satisfaction of the Administration and in accordance with the relevant operational requirements contained in the Guidelines for Inert Gas Systems*.

4 Tankers carrying petroleum products having a flashpoint exceeding 60°C (closed cup test) as determined by an approved flashpoint apparatus shall comply with the provisions of Part C, except that in lieu of the fixed fire-extinguishing system required in regulation 53 they shall be fitted with a fixed deck foam system which shall comply with the provisions of regulation 61.

5 The requirements for inert gas systems of regulation 60 need not be applied to all chemical tankers or gas carriers when carrying cargoes described in paragraph 1, provided that alternative arrangements, to be developed by the Organization, are fitted.**

6 Chemical tankers and gas carriers shall comply with the requirements of this Part, except where alternative and supplementary arrangements are provided to the satisfaction of the Administration, having due regard to the provisions of the Bulk Chemical Code and Gas Carrier Code.

* Reference is made to Guidelines for Inert Gas Systems, adopted by the Maritime Safety Committee at its forty-second session in May 1980 (MSC/Circ.282).

** Reference is made to Interim Regulation for Inert Gas Systems on Chemical Tankers Carrying Petroleum Products, adopted by the Organization by resolution A.473(XII).

Regulation 60

Cargo tank protection

1 For tankers of 20,000 tonnes deadweight and upwards the protection of the cargo tanks deck area and cargo tanks shall be achieved by a fixed deck foam system and a fixed inert gas system in accordance with the requirements of regulations 61 and 62, except that, in lieu of the above installations, the Administration, after having given consideration to the ship's arrangement and equipment, may accept other combinations of fixed installations if they afford protection equivalent to the above, in accordance with regulation I/5.

2 To be considered equivalent, the system proposed in lieu of the deck foam system shall:

.1 be capable of extinguishing spill fires and also preclude ignition of spilled oil not yet ignited; and

.2 be capable of combating fires in ruptured tanks.

3 To be considered equivalent, the system proposed in lieu of the fixed inert gas system shall:

.1 be capable of preventing dangerous accumulations of explosive mixtures in intact cargo tanks during normal service throughout the ballast voyage and necessary in-tank operations; and

.2 be so designed as to minimize the risk of ignition from the generation of static electricity by the system itself.

4 Tankers of 20,000 tonnes deadweight and upwards constructed before 1 September 1984 which are engaged in the trade of carrying crude oil shall be fitted with an inert gas system, complying with the requirements of paragraph 1, not later than:

.1 for a tanker of 70,000 tonnes deadweight and upwards 1 September 1984 or the date of delivery of the ship, whichever occurs later; and

.2 for a tanker of less than 70,000 tonnes deadweight 1 May 1985 or the date of delivery of the ship, whichever occurs later except that for tankers of less than 40,000 tonnes deadweight not fitted with tank washing machines having an individual throughput of greater than 60 m^3/hour the Administration may exempt such tankers from the requirements of this paragraph, if it would be reasonable and impracticable to apply these requirements, taking into account the ship's design characteristics.

5 Tankers of 40,000 tonnes deadweight and upwards constructed before 1 September 1984 which are engaged in the trade of carrying oil other than crude oil and any such tanker of 20,000 tonnes deadweight and upwards engaged in the trade of carrying oil other than crude oil fitted with tank washing machines having an individual throughput of greater than 60 m^3/hour shall be fitted with an inert gas system, complying with the requirements of paragraph 1, not later than:

.1 for a tanker of 70,000 tonnes deadweight and upwards 1 September 1984 or the date of delivery of the ship, whichever occurs later; and

.2 for a tanker of less than 70,000 tonnes deadweight 1 May 1985 or the date of delivery of the ship, whichever occurs later.

6 All tankers operating with a cargo tank cleaning procedure using crude oil washing shall be fitted with an inert gas system complying with the requirements of regulation 62 and with fixed tank washing machines.

7 All tankers fitted with a fixed inert gas system shall be provided with a closed ullage system.

8 Tankers of less than 20,000 tonnes deadweight shall be provided with a deck foam system complying with the requirements of regulation 61.

4. 1983 Amendments to the 1974 SOLAS Convention

(Entry into force: 1 July 1986)

CHAPTER II-2

CONSTRUCTION – FIRE PROTECTION, FIRE DETECTION AND FIRE EXTINCTION

Chapter II-2 of the Convention is replaced by the text of chapter II-2 annexed to resolution MSC.1(XLV), further amended as follows:

Regulation 1

Application

In paragraph 1.1 line 3 delete "1 September 1984" *and insert* "1 July 1986".

In paragraph 1.3.2 line 2 delete "1 September 1984" *and insert* "1 July 1986".

Replace the whole of paragraph 2 by:

"Unless expressly provided otherwise, for ships constructed before 1 July 1986 the Administration shall ensure that the requirements which are applicable under chapter II-2 of the International Convention for the Safety of Life at Sea, 1974, as amended by resolution MSC.1(XLV) are complied with."

In paragraph 3 lines 4 and 9 delete "1 September 1984" *and insert* "1 July 1986".

Delete the footnote.

Regulation 55

Application

Amend paragraph 2 to read:

"Where liquid cargoes other than those referred to in paragraph 1 or liquefied gases which introduce additional fire hazards are intended to be carried, additional safety measures shall be required to the satisfaction of the Administration, having due regard to the provisions of the International Bulk Chemical Code, the Bulk Chemical Code, the International Gas Carrier Code and the Gas Carrier Code, as appropriate."

Amend paragraph 6 to read:

"Chemical tankers and gas carriers shall comply with the requirements of this part, except where alternative and supplementary arrangements are provided to the satisfaction of the Administration, having due regard to the provisions of the International Bulk Chemical Code, the Bulk Chemical Code, the International Gas Carrier Code and the Gas Carrier Code, as appropriate."

5. *Clarification of inert gas system requirements under SOLAS 1974, as amended (MSC/Circ.485)*

1. The Maritime Safety Committee at its fifty-fifth session considered the request for clarification with respect to inert gas system requirements for tankers between 20,000 and 40,000 tonnes deadweight constructed before 1 September 1984.

2. In clarifying this matter the Committee noted that regulation II-2/1.2 of SOLAS 74 as amended by the 1983 amendments states that "... the Administration shall ensure that the requirements which are applicable under chapter II-2 of SOLAS 74, as amended by resolution MSC.1(XLV) adopted on 20 November 1981, are complied with". When reviewing resolution MSC.1(XLV), the 1981 SOLAS amendments, regulation II-2/1.2.2 states that "... for tankers constructed before 1 September 1984, the Administration shall ensure that the requirements which are applicable under chapter II-2 of the Annex to the Protocol of 1978 relating to SOLAS 74, to new or existing ships as defined in that chapter are complied with". Therefore the Committee agreed that the installation of the inert gas system on tankers between 20,000 and 40,000 tonnes deadweight constructed before 1 September 1984 is governed by the requirements of the 1978 SOLAS Protocol.

3. In this regard, regulation II-2/60 of the 1978 SOLAS Protocol requires all tankers (i.e. tankers regardless of deadweight and date of construction) using a crude oil washing system to clean cargo tanks and tankers of 20,000 tonnes deadweight and upwards contracted for after 1 June 1979 or delivered after 1 June 1982 to have an inert gas system installed. For tankers contracted for and delivered before the above dates, the 1978 SOLAS Protocol requires an inert gas system on all such tankers of 20,000 tonnes deadweight and upwards except that:

.1 A tanker of 20,000 tonnes deadweight and above but less than 40,000 tonnes deadweight carrying crude oil is not required to have an inert gas system:

.1.1 if it is not fitted with tank washing machines having an individual throughput of greater than 60 cubic metres per hour; and

.1.2 if the Administration determines it is unreasonable and impracticable, due to the ship's design characteristics, to install an inert gas system;

.2 A tanker of 20,000 tonnes deadweight and above but less than 40,000 tonnes deadweight carrying oil other than crude oil is not required to have an inert gas system if it is not fitted with tank washing machines having an individual throughput of greater than 60 cubic metres per hour.

4. The Committee noted that the actual compliance dates in the 1978 SOLAS Protocol differ from the actual compliance dates in SOLAS 74, as amended in 1981. However, all of these compliance dates have passed and an inert gas system would be required on tankers as stated above.

5. In view of the above clarification, the Committee urged all Administrations to ensure consistent application of the inert gas system requirements as stated above.

PART III PROVISIONS OF TECHNICAL REQUIREMENTS

1. 1974 SOLAS Convention
(Entry into force: 25 May 1980)

Regulation 62

Inert Gas System

The inert gas system referred to in paragraph (a) of regulation 60 of this chapter shall be capable of providing on demand a gas or mixture of gases to the cargo tanks so deficient in oxygen that the atmosphere within a tank may be rendered inert, i.e. incapable of propagating flame. Such a system shall satisfy the following conditions:

(a) The need for fresh air to enter a tank during normal operations shall be eliminated, except when preparing a tank for entry by personnel.

(b) Empty tanks shall be capable of being purged with inert gas to reduce the hydrocarbon content of a tank after discharge of cargo.

(c) The washing of tanks shall be capable of being carried out in an inert atmosphere.

(d) During cargo discharge, the system shall be such as to ensure that the volume of gas referred to in paragraph (f) of this regulation is available. At other times sufficient gas to ensure compliance with paragraph (g) of this regulation shall be continuously available.

(e) Suitable means for purging the tanks with fresh air as well as with inert gas shall be provided.

(f) The system shall be capable of supplying inert gas at a rate of at least 125 per cent of the maximum rated capacity of the cargo pumps.

(g) Under normal running conditions, when tanks are being filled or have been filled with inert gas, a positive pressure shall be capable of being maintained at the tank.

(h) Exhaust gas outlets for purging shall be suitably located in the open air and shall be to the same general requirements as prescribed for ventilating outlets of tanks, referred to in paragraph (a) of regulation 58 of this chapter.

(i) A scrubber shall be provided which will effectively cool the gas and remove solids and sulphur combustion products.

(j) At least two fans (blowers) shall be provided which together shall be capable of delivering at least the amount of gas stipulated in paragraph (f) of this regulation.

(k) The oxygen content in the inert gas supply shall not normally exceed 5 per cent by volume.

(l) Means shall be provided to prevent the return of hydrocarbon gases or vapours from the tanks to the machinery spaces and uptakes and prevent the development of excessive pressure or vacuum. In addition, an effective water lock shall be installed at the scrubber or on deck. Branch piping for inert gas shall be fitted with stop valves or equivalent means of control at every tank. The system shall be so designed as to minimize the risk of ignition from the generation of static electricity.

(m) Instrumentation shall be fitted for continuously indicating and permanently recording at all times when inert gas is being supplied the pressure and oxygen content of the gas in the inert gas supply main on the discharge side of the fan. Such instrumentation should preferably be placed in the cargo control room if fitted but in any case shall be easily accessible to the officer in charge of cargo operations. Portable instruments suitable for measuring oxygen and hydrocarbon gases or vapour and the necessary tank fittings shall be provided for monitoring the tank contents.

(n) Means for indicating the temperature and pressure of the inert gas main shall be provided.

(o) Alarms shall be provided to indicate:

- (i) high oxygen content of gas in the inert gas main;
- (ii) low gas pressure in the inert gas main;
- (iii) low pressure in the supply to the deck water seal, if such equipment is installed;
- (iv) high temperature of gas in the inert gas main; and
- (v) low water pressure to the scrubber

and automatic shutdowns of the system shall be arranged on predetermined limits being reached in respect of subparagraphs (iii), (iv) and (v) of this paragraph.

(p) The master of any ship equipped with an inert gas system shall be provided with an instruction manual covering operational, safety and occupational health requirements relevant to the system.

2. *1981 Amendments to the 1974 SOLAS Convention*
(Entry into force: 1 September 1984)

Regulation 62

Inert gas systems

1 The inert gas system referred to in regulation 60 shall be designed, constructed and tested to the satisfaction of the Administration. It shall be so designed and operated as to render and maintain the atmosphere of the cargo tanks* non-flammable at all times, except when such tanks are required to be gas-free. In the event that the inert gas system is unable to meet the operational requirement set out above and it has been assessed that it is impractical to effect a repair, then cargo discharge, deballasting and necessary tank cleaning shall only be resumed when the "emergency conditions" laid down in the Guidelines on Inert Gas Systems** are complied with.

2 The system shall be capable of:

.1 inerting empty cargo tanks by reducing the oxygen content of the atmosphere in each tank to a level at which combustion cannot be supported;

.2 maintaining the atmosphere in any part of any cargo tank with an oxygen content not exceeding 8 per cent by volume and at a positive pressure at all times in port and at sea except when it is necessary for such a tank to be gas-free;

.3 eliminating the need for air to enter a tank during normal operations except when it is necessary for such a tank to be gas-free;

.4 purging empty cargo tanks of hydrocarbon gas, so that subsequent gas-freeing operations will at no time create a flammable atmosphere within the tank.

3.1 The system shall be capable of delivering inert gas to the cargo tanks at a rate of at least 125 per cent of the maximum rate of discharge capacity of the ship expressed as a volume.

3.2 The system shall be capable of delivering inert gas with an oxygen content of not more than 5 per cent by volume in the inert gas supply main to the cargo tanks at any required rate of flow.

---

* Throughout this regulation the term "cargo tank" includes also "slop tanks".

** Reference is made to Guidelines for Inert Gas Systems, adopted by the Maritime Safety Committee at its forty-second session in May 1980 (MSC/Circ.282).

4 The inert gas supply may be treated flue gas from main or auxiliary boilers. The Administration may accept systems using flue gases from one or more separate gas generators or other sources or any combination thereof, provided that an equivalent standard of safety is achieved. Such systems should, as far as practicable, comply with the requirements of this regulation. Systems using stored carbon dioxide shall not be permitted unless the Administration is satisfied that the risk of ignition from generation of static electricity by the system itself is minimized.

5 Flue gas isolating valves shall be fitted in the inert gas supply mains between the boiler uptakes and the flue gas scrubber. These valves shall be provided with indicators to show whether they are open or shut, and precautions shall be taken to maintain them gastight and keep the seatings clear of soot. Arrangements shall be made to ensure that boiler soot blowers cannot be operated when the corresponding flue gas valve is open.

6.1 A flue gas scrubber shall be fitted which will effectively cool the volume of gas specified in paragraph 3 and remove solids and sulphur combustion products. The cooling water arrangements shall be such that an adequate supply of water will always be available without interfering with any essential services on the ship. Provision shall also be made for an alternative supply of cooling water.

6.2 Filters or equivalent devices shall be fitted to minimize the amount of water carried over to the inert gas blowers.

6.3 The scrubber shall be located aft of all cargo tanks, cargo pump-rooms and cofferdams separating these spaces from machinery spaces of category A.

7.1 At least two blowers shall be fitted which together shall be capable of delivering to the cargo tanks at least the volume of gas required by paragraph 3. In the system with gas generator the Administration may permit only one blower if that system is capable of delivering the total volume of gas required by paragraph 3 to the protected cargo tanks, provided that sufficient spares for the blower and its prime mover are carried on board to enable any failure of the blower and its prime mover to be rectified by the ship's crew.

7.2 Two fuel oil pumps shall be fitted to the inert gas generator. The Administration may permit only one fuel oil pump on condition that sufficient spares for the fuel oil pump and its prime mover are carried on board to enable any failure of the fuel oil pump and its prime mover to be rectified by the ship's crew.

7.3 The inert gas system shall be so designed that the maximum pressure which it can exert on any cargo tank will not exceed the test pressure of any cargo tank. Suitable shutoff arrangements shall be provided on the suction and discharge connections of each blower. Arrangements shall be provided to enable the functioning of the inert gas plant to be stabilized before commencing cargo discharge. If the blowers are to be used for gas-freeing, their air inlets shall be provided with blanking arrangements.

7.4 The blowers shall be located aft of all cargo tanks, cargo pump-rooms and cofferdams separating these spaces from machinery spaces of category A.

8.1 Special consideration shall be given to the design and location of scrubber and blowers with relevant piping and fittings in order to prevent flue gas leakages into enclosed spaces.

8.2 To permit safe maintenance, an additional water seal or other effective means of preventing flue gas leakage shall be fitted between the flue gas isolating valves and scrubber or incorporated in the gas entry to the scrubber.

9.1 A gas regulating valve shall be fitted in the inert gas supply main. This valve shall be automatically controlled to close as required in paragraphs 19.2 and 19.3. It shall also be capable of automatically regulating the flow of inert gas to the cargo tanks unless means are provided to automatically control the speed of the inert gas blowers required in paragraph 7.

9.2 The valve referred to in paragraph 9.1 shall be located at the forward bulkhead of the forwardmost gas-safe space* through which the inert gas supply main passes.

10.1 At least two nonreturn devices, one of which shall be a water seal, shall be fitted in the inert gas supply main, in order to prevent the return of hydrocarbon vapour to the machinery space uptakes or to any gas-safe spaces under all normal conditions of trim, list and motion of the ship. They shall be located between the automatic valve required by paragraph 9.1 and the aftermost connection to any cargo tank or cargo pipeline.

10.2 The devices referred to in paragraph 10.1 shall be located in the cargo tank area on deck.

10.3 The water seal referred to in paragraph 10.1 shall be capable of being supplied by two separate pumps, each of which shall be capable of maintaining an adequate supply at all times.

10.4 The arrangement of the seal and its associated fittings shall be such that it will prevent backflow of hydrocarbon vapours and will ensure the proper functioning of the seal under operating conditions.

10.5 Provision shall be made to ensure that the water seal is protected against freezing, in such a way that the integrity of seal is not impaired by overheating.

10.6 A water loop or other approved arrangement shall also be fitted to each associated water supply and drain pipe and each venting or pressure-sensing pipe leading to gas-safe spaces. Means shall be provided to prevent such loops from being emptied by vacuum.

10.7 The deck water seal and all loop arrangements shall be capable of preventing return of hydrocarbon vapours at a pressure equal to the test pressure of the cargo tanks.

---

* Gas-safe space is a space in which the entry of hydrocarbon gases would produce hazards with regard to flammability or toxicity.

10.8 The second device shall be a nonreturn valve or equivalent capable of preventing the return of vapours or liquids and fitted forward of the deck water seal required in paragraph 10.1. It shall be provided with positive means of closure. As an alternative to positive means of closure, an additional valve having such means of closure may be provided forward of the nonreturn valve to isolate the deck water seal from the inert gas main to the cargo tanks.

10.9 As an additional safeguard against the possible leakage of hydrocarbon liquids or vapours back from the deck main, means shall be provided to permit this section of the line between the valve having positive means of closure referred to in paragraph 10.8 and the valve referred to in paragraph 9 to be vented in a safe manner when the first of these valves is closed.

11.1 The inert gas main may be divided into two or more branches forward of the nonreturn devices required by paragraph 10.

11.2.1 The inert gas supply mains shall be fitted with branch piping leading to each cargo tank. Branch piping for inert gas shall be fitted with either stop valves or equivalent means of control for isolating each tank. Where stop valves are fitted, they shall be provided with locking arrangements, which shall be under the control of a responsible ship's officer.

11.2.2 In combination carriers, the arrangement to isolate the slop tanks containing oil or oil residues from other tanks shall consist of blank flanges which will remain in position at all times when cargoes other than oil are being carried except as provided for in the relevant section of the Guidelines on Inert Gas Systems.

11.3 Means shall be provided to protect cargo tanks against the effect of overpressure or vacuum caused by thermal variations when the cargo tanks are isolated from the inert gas mains.

11.4 Piping systems shall be so designed as to prevent the accumulation of cargo or water in the pipelines under all normal conditions.

11.5 Suitable arrangements shall be provided to enable the inert gas main to be connected to an external supply of inert gas.

12 The arrangements for the venting of all vapours displaced from the cargo tanks during loading and ballasting shall comply with regulation 59.1 and shall consist of either one or more mast risers, or a number of high velocity vents. The inert gas supply mains may be used for such venting.

13 The arrangements for inerting, purging or gas-freeing of empty tanks as required in paragraph 2 shall be to the satisfaction of the Administration and shall be such that the accumulation of hydrocarbon vapours in pockets formed by the internal structural members in a tank is minimized and that:

- .1 on individual cargo tanks the gas outlet pipe, if fitted, shall be positioned as far as practicable from the inert gas/air inlet and in accordance with regulation 59.1. The inlet of such outlet pipes may be located either at deck level or at not more than 1 m above the bottom of the tank;

.2 the cross-sectional area of such gas outlet pipe referred to in paragraph 13.1 shall be such that an exit velocity of at least 20 m/s can be maintained when any three tanks are being simultaneously supplied with inert gas. Their outlets shall extend not less than 2 m above deck level;

.3 each gas outlet referred to in paragraph 13.2 shall be fitted with suitable blanking arrangements;

.4.1 if a connection is fitted between the inert gas supply mains and the cargo piping system, arrangements shall be made to ensure an effective isolation having regard to the large pressure difference which may exist between the systems. This shall consist of two shutoff valves with an arrangement to vent the space between the valves in a safe manner or an arrangement consisting of a spool-piece with associated blanks;

.4.2 the valve separating the inert gas supply main from the cargo main and which is on the cargo main side shall be a nonreturn valve with a positive means of closure.

14.1 One or more pressure/vacuum-breaking devices shall be provided on the inert gas supply main to prevent the cargo tanks from being subject to:

.1 a positive pressure in excess of the test pressure of the cargo tank if the cargo were to be loaded at the maximum specified rate and all other outlets were left shut; or

.2 a negative pressure in excess of 700 mm water gauge if cargo were to be discharged at the maximum rated capacity of the cargo pumps and the inert gas blowers were to fail.

14.2 The location and design of the devices referred to in paragraph 14.1 shall be in accordance with regulation 59.1.

15 Means shall be provided for continuously indicating the temperature and pressure of the inert gas at the discharge side of the gas blowers, whenever the gas blowers are operating.

16.1 Instrumentation shall be fitted for continuously indicating and permanently recording, when the inert gas is being supplied:

.1 the pressure of the inert gas supply mains forward of the nonreturn devices required by paragraph 10.1; and

.2 the oxygen content of the inert gas in the inert gas supply mains on the discharge side of the gas blowers.

16.2 The devices referred to in paragraph 16.1 shall be placed in the cargo control room where provided. But where no cargo control room is provided, they shall be placed in a position easily accessible to the officer in charge of cargo operations.

16.3 In addition, meters shall be fitted:

.1 in the navigating bridge to indicate at all times the pressure referred to in paragraph 16.1.1 and the pressure in the slop tanks of combination carriers, whenever those tanks are isolated from the inert gas supply main; and

.2 in the machinery control room or in the machinery space to indicate the oxygen content referred to in paragraph 16.1.2.

17 Portable instruments for measuring oxygen and flammable vapour concentration shall be provided. In addition, suitable arrangement shall be made on each cargo tank such that the condition of the tank atmosphere can be determined using these portable instruments.

18 Suitable means shall be provided for the zero and span calibration of both fixed and portable gas concentration measurement instruments, referred to in paragraphs 16 and 17.

19.1 Audible and visual alarms shall be provided to indicate:

.1 low water pressure or low water flow rate to the flue gas scrubber as referred to in paragraph 6.1;

.2 high water level in the flue gas scrubber as referred to in paragraph 6.1;

.3 high gas temperature as referred to in paragraph 15;

.4 failure of the inert gas blowers referred to in paragraph 7;

.5 oxygen content in excess of 8 per cent by volume as referred to in paragraph 16.1.2;

.6 failure of the power supply to the automatic control system for the gas regulating valve and to the indicating devices as referred to in paragraphs 9 and 16.1;

.7 low water level in the water seal as referred to in paragraph 10.1;

.8 gas pressure less than 100 mm water gauge as referred to in paragraph 16.1.1. The alarm arrangement shall be such as to ensure that the pressure in slop tanks in combination carriers can be monitored at all times; and

.9 high gas pressure as referred to in paragraph 16.1.1.

19.2 In the system with gas generators audible and visual alarms shall be provided in accordance with 19.1.1, 19.1.3, 19.1.5 to 19.1.9 and additional alarms to indicate:

.1 insufficient fuel oil supply;

.2 failure of the power supply to the generator;

.3 failure of the power supply to the automatic control system for the generator.

19.3 Automatic shutdown of the inert gas blowers and gas regulating valve shall be arranged on predetermined limits being reached in respect of paragraphs 19.1.1, 19.1.2 and 19.1.3.

19.4 Automatic shutdown of the gas regulating valve shall be arranged in respect of paragraph 19.1.4.

19.5 In respect of paragraph 19.1.5, when the oxygen content of the inert gas exceeds 8 per cent by volume, immediate action shall be taken to improve the gas quality. Unless the quality of the gas improves, all cargo tank operations shall be suspended so as to avoid air being drawn in to the tanks and the isolation valve referred to in paragraph 10.8 shall be closed.

19.6 The alarms required in paragraphs 19.1.5, 19.1.6 and 19.1.8 shall be fitted in the machinery space and cargo control room, where provided, but in each case in such a position that they are immediately received by responsible members of the crew.

19.7 In respect of paragraph 19.1.7 the Administration shall be satisfied as to the maintenance of an adequate reserve of water at all times and the integrity of the arrangements to permit the automatic formation of the water seal when the gas flow ceases. The audible and visual alarm on the low level of water in the water seal shall operate when the inert gas is not being supplied.

19.8 An audible alarm system independent of that required in paragraph 19.1.8 or automatic shutdown of cargo pumps shall be provided to operate on predetermined limits of low pressure in the inert gas mains being reached.

20 Tankers constructed before 1 September 1984 which are required to have an inert gas system shall at least comply with the requirements of regulation 62 of chapter II-2 of the International Convention for the Safety of Life at Sea, 1974*. In addition they shall comply with the requirements of this regulation, except that:

.1 inert gas systems fitted on board such tankers before 1 June 1981 need not comply with the following paragraphs: 3.2, 6.3, 7.4, 8, 9.2, 10.2, 10.7, 10.9, 11.3, 11.4, 13.2, 13.4.2 and 19.8;

.2 inert gas systems fitted on board such tankers on or after 1 June 1981 need not comply with the following paragraphs: 3.2, 6.3, 7.4 and 13.2.

21 Detailed instruction manuals shall be provided on board, covering the operations, safety and maintenance requirements and occupational health hazards relevant to the inert gas system and its application to the cargo tank system**. The manuals shall include guidance on procedures to be followed in the event of a fault or failure of the inert gas system.

---

* The text as adopted by the International Conference on Safety of Life at Sea, 1974.

** Reference is made to Guidelines for Inert Gas Systems, adopted by the Maritime Safety Committee at its forty-second session in May 1980 (MSC/Circ.282).

3. 1983 Amendments to the 1974 SOLAS Convention
(Entry into force: 1 July 1986)

Regulation 62

Inert gas systems

In paragraph 1 delete "non fllammable" *and insert* "non flammable".

In paragraph 9.1, lines 2 and 3 delete "19.2" *and* "19.3" *and insert* "19.3" *and* "19.4" *respectively.*

In paragraph 10.2 amend "cargo tank area" *to read* "cargo area".

Replace paragraph 14.1 by:

"14.1 One or more pressure/vacuum-breaking devices shall be provided to prevent the cargo tanks from being subject to:

.1 a positive pressure in excess of the test pressure of the cargo tank if the cargo were to be loaded at the maximum rated capacity and all other outlets are left shut; and

.2 a negative pressure in excess of 700 mm water gauge if cargo were to be discharged at the maximum rated capacity of the cargo pumps and the inert gas blowers were to fail.

Such devices shall be installed on the inert gas main unless they are installed in the venting system required by regulation 59.1.1 or on individual cargo tanks."

In paragraph 20.1 amend the last line to read "10.2, 10.7, 10.9, 11.3, 11.4, 12, 13.1, 13.2, 13.4.2, 14.2 and 19.8;"

In paragraph 20.2 amend the last line to read "12, 13.1, 13.2 and 14.2."

PART IV REGULATIONS FOR INERT GAS SYSTEMS ON CHEMICAL TANKERS

1. Resolution A.567(14)

(adopted on 20 November 1985 by the IMO Assembly at its fourteenth session)

REGULATION FOR INERT GAS SYSTEMS ON CHEMICAL TANKERS

THE ASSEMBLY,

RECALLING Article 15(j) of the Convention on the International Maritime Organization concerning the functions of the Assembly in relation to regulations and guidelines concerning maritime safety,

RECALLING ALSO resolution A.473(XII) which was adopted to provide an interim solution for the requirements of inert gas systems applicable to chemical tankers carrying petroleum products, pending the possible development of final requirements applicable to chemical tankers carrying all flammable cargoes,

RECOGNIZING that the development of such requirements is not needed on the basis of results of scientific studies undertaken by industry, but that the extension of the regulation in resolution A.473(XII) to cover the carriage of petroleum and other liquid products would meet the purpose,

NOTING that regulation II-2/60 of the International Convention for the Safety of Life at Sea, 1974 (1974 SOLAS Convention) as amended requires *inter alia* new and existing tankers of a certain size, including chemical tankers, when carrying petroleum products, to be fitted with a fixed inert gas system by specific dates,

NOTING FURTHER that the draft amendment to regulation II-2/55.5 of the 1974 SOLAS Convention as amended (resolution A.566(14)) exempts certain chemical tankers and gas carriers carrying flammable products from the requirements for inert gas systems of regulation II-2/60 of that Convention under certain conditions,

HAVING CONSIDERED the recommendation made by the Maritime Safety Committee at its fifty-first session,

1. ADOPTS the Regulation for Inert Gas Systems on Chemical Tankers set out in the Annex to the present resolution, which supersedes resolution A.473(XII);

2. INVITES Governments to apply the above regulation to chemical tankers for the purpose of the implementation of the draft amendment to regulation II-2/55.5.

ANNEX

REGULATION FOR INERT GAS SYSTEMS ON CHEMICAL TANKERS

PREAMBLE

Administrations are invited to accept the inert gas systems referred to in this regulation for chemical tankers for which certificates of fitness are issued under the Code for the Construction and Equipment of Ships Carrying Dangerous Chemicals in Bulk (resolution A.212(VII)) and under the International Code for the Construction and Equipment of Ships Carrying Dangerous Chemicals in Bulk (resolution MSC.4(48)).

This regulation shall be applied to chemical tankers as required by the draft amendment to regulation II-2/55.5 of the 1974 SOLAS Convention as amended (resolution A.566(14)).

REGULATION

1 Inert gas generator systems* shall be designed, constructed and tested to the satisfaction of the Administration. They shall be designed and operated so as to render and maintain the atmosphere of cargo tanks** non-flammable at all times except when such tanks are required to be maintained empty and gas-free. Inert gas systems supplied by one or more oil-fired inert gas generators may be accepted. An Administration may accept systems using inert gas from other sources provided that an equivalent standard of safety is achieved.

2 The systems shall be capable of:

.1 inerting empty cargo tanks by reducing the oxygen content of the atmosphere in each tank to a level at which combustion cannot be supported;

.2 maintaining the atmosphere, in all parts of each cargo tank designated to carry flammable products requiring protection by an inert gas system, with an oxygen content not exceeding 8% by volume and at a positive pressure at all times in port and at sea except when it is necessary for such a tank to be gas-free;

.3 eliminating the need for air to enter a tank during normal operations except when it is necessary for such a tank to be gas-free;

---

* "Inert gas generator system" means the machinery dedicated to the production and supply of inert gas and includes the air blowers, combustion chambers, oil fuel pumps and burners, gas coolers/scrubbers and automatic combustion control and supervisory equipment, e.g. flame failure devices.

** Throughout this regulation the term "cargo tank" includes also "slop tanks containing oil residues".

.4 purging empty cargo tanks of flammable vapour, so that subsequent gas-freeing operations will at no time create a flammable atmosphere within the tank.

3.1 The systems shall be capable of delivering inert gas to the cargo tanks at a rate of at least 125% of the maximum rate of discharge capacity of the ship expressed as a volume. An Administration may accept inert gas systems having a lower delivery capacity provided that the maximum rate of discharge of cargoes from cargo tanks being protected by the system is restricted to 80% of the inert gas capacity.

3.2 The systems shall be capable of delivering inert gas with an oxygen content of not more than 5% by volume in the inert gas supply main to the cargo tanks at any required rate of flow.

4.1 Suitable fuel in sufficient quantity shall be provided for the inert gas generators.

4.2 The inert gas generators shall be located outside the cargo tank area as defined in the Bulk Chemical Code and the International Bulk Chemical Code. Spaces containing inert gas generators should have no direct access to accommodation, service or control station spaces, but may be located in machinery spaces. If they are not located in machinery spaces they shall be located in a compartment reserved solely for their use. Such a compartment shall be separated by a gastight steel bulkhead and/or deck from accommodation, service and control station spaces as defined in the Bulk Chemical Code and the International Bulk Chemical Code. Adequate positive-pressure-type mechanical ventilation shall be provided for such a compartment. Access to such compartments located aft shall be only from an open deck outside the cargo tank area. Access shall be located on the end bulkhead not facing the cargo area and/or on the outboard side of the superstructure or deckhouse at a distance of at least 25% of the length of the ship but not less than 5 m from the end of the superstructure or deckhouse facing the cargo area. In the case of such a compartment being located in the forecastle, access shall be through the deckhead forward of the cargo area.

4.3 Inert gas piping systems shall not pass through accommodation, service and control station spaces.

5.1 Means shall be provided which will effectively cool the volume of gas specified in paragraph 3 and remove solids and sulphur combustion products. The cooling water arrangements shall be such that an adequate supply of water will always be available without interfering with any essential services on the ship. Provision shall also be made for an alternative supply of cooling water.

5.2 Filters or equivalent devices shall be fitted to minimize the amount of water carried over to the inert gas main.

6.1 Two air blowers shall be fitted to each inert gas generator, which together shall be capable of delivering to the cargo tanks, required to be protected by the system, at least the volume of gas required by paragraph 3. An Administration may permit only one blower if it is capable of delivering to the protected cargo tanks the total volume of gas required by paragraph 3, provided that sufficient spares for the air blower and its prime mover are carried on board to enable any failure of the air blower and its prime mover to be rectified.

6.2 The inert gas systems shall be so designed that the maximum pressure which they can exert on any cargo tank will not exceed the test pressure of any cargo tank.

6.3 Where more than one inert gas generator is provided, suitable shutoff arrangements shall be provided on the discharge outlet of each generator plant.

6.4 Arrangements shall be made to vent the inert gas to the atmosphere in case the inert gas produced is off-specification, e.g. during starting-up or in case of equipment failure.

6.5 Where inert gas generators are served by positive displacement blowers, a pressure relief device shall be provided to prevent excess pressure being developed on the discharge side of the blower.

7 Two fuel oil pumps shall be fitted to each inert gas generator. An Administration may permit only one fuel oil pump on condition that sufficient spares for the fuel oil pump and its prime mover are carried on board to enable any failure of the fuel oil pump and its prime mover to be rectified by the ship's crew.

8 A gas regulating valve shall be fitted in the inert gas supply main. This valve shall be automatically controlled to close as required in paragraphs 17.2 and 17.3. It shall also be capable of automatically regulating the flow of inert gas to the cargo tanks unless other means are provided to automatically control the inert gas flow rate.

9.1 At least two nonreturn devices, one of which shall be a water seal, shall be fitted in the inert gas supply main in order to prevent the return of flammable vapour to the inert gas generator and to any gas-safe space under all normal conditions of trim, list and motion of the ship. They shall be located between the automatic valve required by paragraph 8 and the first connection to any cargo tank or cargo pipeline. An Administration may permit an alternative arrangement or device providing a measure of safety equivalent to that of a water seal.

9.2 The devices referred to in paragraph 9.1 shall be located in the cargo tank area on deck.

9.3 The water seal referred to in paragraph 9.1 shall be capable of being supplied by two separate pumps, each of which shall be capable of maintaining an adequate supply at all times.

9.4 The arrangement of the water seal and its associated provisions shall be such that it will prevent backflow of flammable vapours and will ensure the proper functioning of the water seal under operating conditions.

9.5 Provisions shall be made to ensure that any water seal is protected against freezing, in such a way that the integrity of water seal is not impaired by overheating.

9.6 A water loop or other approved arrangement shall also be fitted to all associated water supply and drain piping and to all venting or pressure sensing piping leading to gas-safe spaces.* Means shall be provided to prevent such loops from being emptied by vacuum.

9.7 Any water seal or equivalent device and all loop arrangements shall be capable of preventing the return of flammable vapours to an inert gas generator at a pressure equal to the test pressure of the cargo tanks.

9.8 The second device shall be a nonreturn valve or equivalent capable of preventing the return of vapours or liquids or both and fitted between the water seal (or the equivalent device) required in paragraph 9.1 and the first connection from the inert gas main to a cargo tank. It shall be provided with positive means of closure. As an alternative to positive means of closure, an additional valve having such means of closure may be provided between the nonreturn valve and the first connection to the cargo tanks to isolate the water seal (or equivalent device).

9.9 As an additional safeguard against the possible leakage of flammable liquids or vapours back from the deck main, means shall be provided to permit this section of the line between the valve having positive means of closure referred to in paragraph 9.8 and the valve referred to in paragraph 8 to be vented in a safe manner when the first of these valves is closed.

10.1 The inert gas main may be divided into two or more branches between the nonreturn devices required by paragraph 9 and the cargo tanks.

10.2 Inert gas supply mains shall be fitted with branch piping leading to each cargo tank designated for the carriage of flammable products required to be inerted by this regulation. Each cargo tank containing or loading products not required to be inerted shall be separated from the inert gas main by:

- .1 removing spool-pieces, valves or other pipe sections, and blanking the pipe ends; or
- .2 arrangement of two spectacle flanges in series with provisions for detecting leakage into the pipe between the two spectacle flanges.

10.3 Means shall be provided to protect cargo tanks against the effect of overpressure or vacuum caused by thermal variations when the cargo tanks are isolated from the inert gas mains.

10.4 Piping systems shall be so designed as to prevent the accumulation of cargo or water in the pipelines under all normal conditions.

10.5 Suitable arrangements shall be provided to enable the inert gas main to be connected to an external supply of inert gas.

* Gas-safe space is a space in which the entry of hydrocarbon gases would produce hazards with regard to flammability or toxicity.

11 Unless the arrangements for venting of all vapours displaced from the cargo tanks during loading and ballasting comply with the requirements of the BCH and IBC Codes for controlled venting, such arrangements shall comply with regulation II-2/59.1 of SOLAS 1974 as amended and shall consist either of one or more mast risers or of a number of high velocity vents.

12 The arrangements for inerting, purging or gas-freeing of empty tanks as required in paragraph 2 shall be to the satisfaction of the Administration and shall be such that the accumulation of hydrocarbon vapours in pockets formed by the internal structural members in a tank is minimized and that:

.1 on individual cargo tanks the gas outlet pipe, if fitted, shall be positioned as far as practicable from the inert gas/air inlet and in accordance with regulation II-2/59.1.9.3 of the 1974 SOLAS Convention as amended, or 8.2.2.3 of the IBC Code. The inlet of such outlet pipes may be located either at deck level or at not more than 1 m above the bottom of the tank;

.2 the cross-sectional area of such gas outlet pipe referred to in subparagraph 12.1 shall be such that an exit velocity of at least 20 m/s can be maintained when any three tanks are being simultaneously supplied with inert gas. Their outlets shall extend not less than 2 m above deck level. When in accordance with paragraph 3 an Administration permits a system designed to supply only one or two tanks simultaneously, the outlet pipes should be sized such that an exit velocity in the outlet pipes of 20 m/s can be maintained;

.3 each gas outlet referred to in subparagraph 12.2 shall be fitted with suitable blanking arrangements.

13 Means shall be provided for continuously indicating the temperature and pressure of the inert gas at the discharge side of the system, whenever it is operating.

14.1 Instrumentation shall be fitted for continuously indicating and permanently recording, when the inert gas is being supplied:

.1 the pressure of the inert gas supply mains between the nonreturn devices required by paragraph 9.1 and the cargo tanks; and

.2 the oxygen content of the inert gas in the inert gas supply main.

14.2 The devices referred to in paragraph 14.1 shall be placed in the cargo control room where provided. Where no cargo control room is provided, they shall be placed in a position easily accessible to the officer in charge of cargo operations.

14.3 In addition, meters shall be fitted:

.1 in the navigating bridge to indicate at all times the pressure referred to in paragraph 14.1.1; and

.2 in the machinery control room or in the machinery space to indicate the oxygen content referred to in paragraph 14.1.2.

15 Portable instruments for measuring oxygen and flammable vapour concentration shall be provided. In addition, suitable arrangement shall be made on each cargo tank such that the condition of the tank atmosphere can be determined using these portable instruments.

16 Suitable means shall be provided for the zero and span calibration of both fixed and portable gas concentration measurement instruments, referred to in paragraphs 14 and 15.

17.1 Audible and visual alarms shall be provided to indicate:

.1 low water pressure or low water flow rate to the cooling and scrubbing arrangement referred to in paragraph 5.1;

.2 low fuel supply;

.3 high gas temperature as referred to in paragraph 13;

.4 failure of the power supply to the inert gas generators;

.5 oxygen content in excess of 8% by volume as referred to in paragraph 14.1.2;

.6 failure of the power supply to the indicating devices as referred to in paragraph 14.1 and to the automatic control systems for the gas regulating valve referred to in paragraph 8 and the inert gas generator;

.7 low water level in the water seal as referred to in paragraph 9.1;

.8 gas pressure less than 100 mm water gauge as referred to in paragraph 14.1;

.9 high gas pressure as referred to in paragraph 14.1.1.

17.2 Automatic shutdown of the gas regulating valve and of the fuel oil supply to the inert gas generator shall be arranged on predetermined limits being reached in respect of paragraphs 17.1.1 and 17.1.3.

17.3 Automatic shutdown of the gas regulating valve shall be arranged in respect of paragraph 17.1.4.

17.4 In respect of paragraph 17.1.5, when the oxygen content of the inert gas exceeds 8% by volume, immediate action shall be taken to improve the gas quality. Unless the quality of the inert gas improves, all operations in those tanks to which inert gas is being supplied shall be suspended so as to avoid air being drawn into the tanks. The deck isolation valve referred to in paragraph 9.8 shall be closed and the off-specification gas shall be vented to atmosphere.

17.5 The alarms required in paragraphs 17.1.5, 17.1.6 and 17.1.8 shall be fitted in the machinery space and cargo control room, where provided, but in each case in such a position that they are immediately received by responsible members of the crew. All other alarms required by this paragraph shall be audible to responsible members of the crew either as individual alarms or as a group alarm.

17.6 In respect of paragraph 17.1.7 the Administration shall be satisfied as to the maintenance of an adequate reserve of water at all times and the integrity of the arrangements to permit the automatic formation of the water seal when the gas flow ceases. The audible and visual alarm on the low level of water in the water seal shall operate when the inert gas is not being supplied.

17.7 An audible alarm system independent of that required in paragraph 17.1.8 or automatic shutdown of cargo pumps shall be provided to operate on predetermined limits of low pressure in the inert gas mains being reached.

18 Detailed instruction manuals shall be provided on board, covering the operations, safety and maintenance requirements and occupational health hazards relevant to the inert gas system and its application to the cargo tank system. The manuals shall include guidance on procedures to be followed in the event of a fault or failure of the inert gas system.

2. *Resolution A.566(14)*

(adopted on 20 November 1985 by the IMO Assembly at its fourteenth session)

DRAFT AMENDMENT TO REGULATION II-2/55.5 OF THE INTERNATIONAL CONVENTION FOR THE SAFETY OF LIFE AT SEA, 1974, AS AMENDED

THE ASSEMBLY,

RECALLING Article 15(j) of the Convention on the International Maritime Organization concerning the functions of the Assembly in relation to regulations and guidelines concerning maritime safety,

BEARING IN MIND regulation I/5 of the International Convention for the Safety of Life at Sea, 1974 (1974 SOLAS Convention), as amended, concerning equivalents,

RECALLING ALSO that regulation II-2/55.5 of the 1974 SOLAS Convention as amended refers to alternative arrangements to the inert gas system requirements of regulation II-2/60 for chemical tankers and gas carriers to be developed by the Organization,

RECALLING FURTHER resolution A.473(XII) which provides an interim regulation for inert gas systems on chemical tankers carrying petroleum products for the purpose of implementation of regulation II-2/55.5,

NOTING that by that resolution the Assembly agreed that compliance with additional provisions which will be contained in the final requirements should not be required to be applied to ships the keels of which are laid before the date of entry into force of the final requirements,

NOTING FURTHER that chemical tankers carrying flammable chemical products constructed hitherto have been allowed to operate without inert gas systems, because of the absence of any regulations applicable to such ships, and that this status may continue to be allowed in future for such ships,

HAVING ADOPTED, by resolution A.567(14), the Regulation for Inert Gas Systems on Chemical Tankers which applies to inert gas systems on chemical tankers carrying petroleum and chemical products,

HAVING CONSIDERED the recommendation made by the Maritime Safety Committee at its fifty-first session,

1. NOTES the draft amendment to regulation II-2/55.5 of the 1974 SOLAS Convention, the text of which is set out in the Annex to the present resolution;

2. REQUESTS the Maritime Safety Committee to consider adoption of the draft amendment to regulation II-2/55.5 at the earliest opportunity;

3. RECOMMENDS that, pending the entry into force of the above amendment, Governments concerned apply the requirements of the draft amendment, as an equivalent to the existing requirements of regulation II-2/55.5 of the 1974 SOLAS Convention as amended.

ANNEX

DRAFT AMENDMENT TO REGULATION II-2/55.5 OF THE INTERNATIONAL CONVENTION FOR THE SAFETY OF LIFE AT SEA, 1974, AS AMENDED

55.5 The requirements for inert gas systems of regulation 60 need not be applied to:

.1 chemical tankers constructed before, on or after 1 July 1986 when carrying cargoes described in paragraph 1, provided that they comply with the requirements for inert gas systems on chemical tankers developed by the Organization*; or

.2 chemical tankers constructed before 1 July 1986, when carrying crude oil or petroleum products, provided that they comply with the requirements for inert gas systems on chemical tankers carrying petroleum products developed by the Organization**; or

.3 gas carriers constructed before, on or after 1 July 1986 when carrying cargoes described in paragraph 1, provided that they are fitted with cargo tank inerting arrangements equivalent to those specified in subparagraph 5.1 or 5.2; or

.4 chemical tankers and gas carriers when carrying flammable cargoes other than crude oil or petroleum products such as cargoes listed in chapters VI and VII of the Code for the Construction and Equipment of Ships Carrying Dangerous Chemicals in Bulk or chapters 17 and 18 of the International Code for the Construction and Equipment of Ships Carrying Dangerous Chemicals in Bulk:

.4.1 if constructed before 1 July 1986; or

---

* Reference is made to Regulation for Inert Gas Systems on Chemical Tankers adopted by the Organization by resolution A.567(14).

** Reference is made to Interim Regulation for Inert Gas Systems on Chemical Tankers Carrying Petroleum Products, adopted by the Organization by resolution A.473(XII).

.4.2 if constructed on or after 1 July 1986, provided that the capacity of tanks used for their carriage does not exceed 3,000 m^3 and the individual nozzle capacities of tank washing machines do not exceed 17.5 m^3/h and the total combined throughput from the number of machines in use in a cargo tank at any one time does not exceed 110 m^3/h.

PART V APPLICATION OF REQUIREMENTS FOR INERT GAS SYSTEMS FOR OIL TANKERS BY PORT AUTHORITIES AND TERMINAL OPERATORS (MSC/Circ.329)

APPLICATION OF REQUIREMENTS FOR INERT GAS SYSTEMS FOR OIL TANKERS

Introduction

At the forty-sixth session of the Maritime Safety Committee, concern was expressed on the stringent oxygen levels insisted on by some terminal operators and port authorities for inerted cargo tanks of oil tankers, and their reluctance to allow the opening of inerted tanks for dipping, measuring and sampling. The Committee, noting the concern, urges Governments to encourage port authorities and terminal operators to comply with international requirements.

Oxygen levels in inerted cargo tanks

Tanker operators have experienced difficulties in certain ports in which the port authorities have introduced more stringent requirements regarding the maximum acceptable oxygen level in inerted cargo tanks than those established by the Organization. The attention of Administrations is therefore drawn to the requirements adopted by the Organization in Conventions and Recommendations.*

The Organization has been informed that some oil tanker terminals have been insisting on oxygen levels in the cargo tanks as low as a maximum of 5% by volume. The regulations and requirements adopted by the Organization regarding oxygen levels in cargo tanks specify a maximum of 8% by volume, and this is considered to give an adequate safety margin for in-tank operations relating to cargo handling, ballasting, and crude oil washing.

Although regulation 62(k) of chapter II-2 of the 1974 SOLAS Convention requires that the oxygen level in the inert gas supply shall not normally exceed 5% by volume, it was never intended that this lower oxygen level should have to be achieved within the cargo tanks.

Administrations will be aware that the enforcement of these more stringent requirements may preclude cargo operations and prevent ships from implementing the MARPOL 73/78 requirement given in Annex I, regulation 13(B)(4), that ballast water shall only be put into cargo tanks which have been crude oil washed.

---

* International Convention for the Safety of Life at Sea, 1974.
International Convention for the Prevention of Pollution from Ships, 1973/78.
1981 Amendments to the International Convention for the Safety of Life at Sea, 1974.
Guidelines for Inert Gas Systems (MSC/Circ.282).
Revised Specifications for the Design, Operation and Control of Crude Oil Washing Systems (resolution A.446(XI)).

Dipping, measuring and sampling of inerted cargo tanks

The Organization has also been informed of a reluctance by certain terminal operators to allow the opening of inerted tanks for dipping, measuring and sampling. These are operations which are necessary in connection with normal cargo handling and safe crude oil washing procedures. These operations are acceptable, provided that a positive inert gas pressure is maintained in the cargo tanks and that the recommended safety precautions* are supplied.

Member Governments are invited to take note of the above difficulties which have been encountered with requirements applied by port authorities for oxygen levels in inerted cargo tanks and opening of cargo ullage ports. They are urged to take whatever steps are possible to ensure that port authorities and terminal operators implement procedures which are consistent with the requirements contained in Conventions and Recommendations adopted by the Organization.

---

* Reference is made to the Guidelines for Inert Gas Systems (MSC/Circ.282) and to chapter 9 of the *International Safety Guide for Oil Tankers and Terminals (ISGOTT).*

NOTES